Tropane Alkaloids

Vikas Srivastava • Shakti Mehrotra •
Sonal Mishra
Editors

Tropane Alkaloids

Pathways, Potential and Biotechnological Applications

Springer

Editors
Vikas Srivastava
Department of Botany
Central University of Jammu
Samba, Jammu and Kashmir, India

Shakti Mehrotra
Department of Biotechnology
Institute of Engineering and Technology
Lucknow, Uttar Pradesh, India

Sonal Mishra
School of Biotechnology
University of Jammu
Jammu, Jammu and Kashmir, India

ISBN 978-981-33-4534-8   ISBN 978-981-33-4535-5 (eBook)
https://doi.org/10.1007/978-981-33-4535-5

This Springer imprint is published by the registered company Springer Nature Singapore Pte Ltd.
The registered company address is: 152 Beach Road, #21-01/04 Gateway East, Singapore 189721, Singapore

This book is dedicated to our
Grandfather Late Shri Diksha Ram Srivastava

Preface

Alkaloids constitute a major group of plant secondary metabolites and have high pharmacological activities. Alkaloids with a tropane ring in their chemical structure or else referred to as tropane alkaloids (TAs) belong to one of the world's oldest plant medicines, with their valuable ethnopharmacological applications. Currently, several aspects are being explored globally regarding plant-based TA production which needs a judicial updation and compilation of existing information, challenges, and future perspectives of the subject. Keeping this in consideration, the book entitled "Tropane Alkaloids: Pathways, Potential, and Biotechnological Applications" has been designed, which covers various features of plant-based TAs, their biosynthetic pathway, and allied applications. The book provides a holistic compilation of scientific ideas, perspectives, and challenges/limitations of this area in the current scenario. One of the purposes of the book is to create a cumulative information repository on various biotechnological interventions in plant-based TAs production. The book addresses critical discussions of subject-related problems, and logical solutions to different reader communities including scientific, academic, and industry based readers. The readers will be benefitted by technical, methodical, and sensible source of the subject information in one compilation, through well-thought-out scientific discussions from various eminent research groups. Entirely, the book is divided into two sections.

Part I (Tropane Alkaloids: Diversity, Biosynthesis, and Significance) comprised of three chapters and includes a focus on chemotaxonomic significance, biogenesis, and biotechnological interventions (Chap. 1), structural and functional details of enzymes involved in the TAs biosynthesis (Chap. 2), and commercial stature and production methods (Chap. 3) of tropane alkaloids.

Part II (Tropane Alkaloids: In Vitro and Allied Interventions) comprised of six chapters and includes a topic on in vitro production of TAs and allied applications of TAs bearing plants (Chap. 4), the contribution of hairy root cultures for TAs production (Chap. 5), TAs production in plant and in vitro cultures of different ploidy levels (Chap. 6), progress in TAs production by transgenic and heterologous gene expression (Chap. 7), insights from transcriptome analysis for TAs biosynthesis (Chap. 8), and transcription factor and microRNA-mediated manipulation of the TAs (Chap. 9).

Precisely, this edited book is an attempt to present an effective asset to those who wish to work on TAs-based research. With this book, we attempt to ensure the research and teaching community about the major progress in TAs bearing plants and their biotechnological explorations. The compilation will surely provide endless opportunities in current and future research in this fascinating area.

Samba, India
Lucknow, India
Jammu, India

Vikas Srivastava
Shakti Mehrotra
Sonal Mishra

Contents

Part I Tropane Alkaloids: Diversity, Biosynthesis and Significance

1 **Plant Tropane Alkaloids (TAs): Chemotaxonomic Significance, Biogenesis, and Biotechnological Interventions** 3
Savita, Anju Srivastava, Reena Jain, Avinash Kaur Nagpal, and Pratap Kumar Pati

2 **Structure and Function of Enzymes Involved in the Biosynthesis of Tropane Alkaloids** 21
Neill Kim, Benjamin Chavez, Charles Stewart Jr., and John C. D'Auria

3 **Plant Tropane Alkaloids: Commercial Stature and Production Developments** 51
Shakti Mehrotra, Sonal Mishra, and Vikas Srivastava

Part II Tropane Alkaloids: In Vitro and Allied Interventions

4 **Involvement of Various Biotechnological Contrivances for Tropane Alkaloid Biosynthesis and Applications of Tropane Alkaloid-Bearing In Vitro Cultures** 69
Vikas Srivastava, Sonal Mishra, Skalzang Lhamo, Aksar Ali Chowdhary, and Shakti Mehrotra

5 **Hairy Root Cultures and Plant Tropane Alkaloids: Production Matrix and Product in Biotechnological Perspective** 89
Shital K. Sharma and Adarsh K. Agnihotri

6 **Production of Tropane Alkaloids (TAs) in Plants and In Vitro Cultures of Different Ploidy Levels** 105
Ishmael Dehghan and Elnaz Ghotbi

7 **Current Progress in Tropane Alkaloid Production by Transgenic and Heterologous Gene Expression Approaches** 113
Neeraj Kumar Dubey, Prashant Singh, Ankita Singh, and Satyendra Kumar Yadav

8 **Tropane Alkaloid Biosynthesis in Plants: Insights from Transcriptome Analysis** 133
Asosii Paul, Bendangchuchang Longchar, and Jeremy Dkhar

9 **Transcription Factor and MicroRNA-Mediated Manipulation of Tropane Alkaloid Biosynthesis** 157
Pravin Prakash, Rakesh Srivastava, and Praveen Chandra Verma

Editors and Contributors

About the Editors

Vikas Srivastava is working as Assistant Professor in the Department of Botany, Central University of Jammu, India. He completed his Ph.D. jointly from Central Institute of Medicinal and Aromatic Plants (CSIR-CIMAP) and Lucknow University, India. Furthermore, he pursued Postdoc from the National Institute of Plant Genome Research (NIPGR), New Delhi, India. With specialization in various aspects of biotechnology, he has acquired training and experience to carry out research in diverse field of plant biology. He has published many articles in journals/books of international repute and has been working as Principal Investigator for major project sanctioned by University Grant Commission (UGC), New Delhi. He is a recipient of various awards and prestigious fellowships. He delivered several invited lectures in institutes of national repute.

Shakti Mehrotra is Consulting Scientist in the Department of Biotechnology, Institute of Engineering and Technology, Dr. A.P.J. Abdul Kalam Technical University, Lucknow, India. She completed her Ph.D. jointly from CSIR-CIMAP and Lucknow University, India. Furthermore, she pursued Postdoc from the Institute of Engineering and Technology, Lucknow (DBT-PDF) and CSIR-CIMAP (DST Young Scientist). With specialization in various aspects of biotechnology, she has acquired training and experience to carry out research in diverse field of plant biotechnology. She has published many articles in journals/books of national and international repute and worked as Principal Investigator for major projects sanctioned by the Department of Science and Technology (DST) and Department of Biotechnology (DBT), Government of India. She is a recipient of various awards and prestigious fellowships. She delivered several presentations in conferences and workshops and received appreciation.

Sonal Mishra is currently working as Dr. D. S. Kothari-Post Doctoral Fellow (UGC-DSKPDF) at the School of Biotechnology, University of Jammu, India. She has recently completed SERB-National Postdoc Fellowship (SERB-NPDF) as Prinicipal Investigator at School of Biotechnology, University of Jammu, India. Further, she completed her Postdoc at Jawaharlal Nehru University and NIPGR, New Delhi, India. She pursued her Ph.D. jointly from CSIR-CIMAP, Lucknow and Jawaharlal Nehru University, New Delhi. With specialization in various aspects of biotechnology, she has acquired training and experience to carry out research in diverse fields of plant biotechnology and molecular biology. She has published many articles in journals/ books of international repute. She is a recipient of various awards and prestigious fellowships. She presented her work in several seminar and conferences and received appreciation.

Contributors

Adarsh K. Agnihotri Bio-processing and Herbal Division, Mahatma Gandhi Institute for Rural Industrialization, Wardha, Maharashtra, India

Benjamin Chavez Department of Molecular Genetics IPK Gatersleben, Leibniz Institute of Plant Genetics and Crop Plant Research, Gatersleben, Germany

Aksar Ali Chowdhary Department of Botany, Central University of Jammu, Samba, Jammu and Kashmir, India

John C. D'Auria Department of Molecular Genetics IPK Gatersleben, Leibniz Institute of Plant Genetics and Crop Plant Research, Gatersleben, Germany

Ishmael Dehghan Department of Molecular Biology, University of Texas Southwestern Medical Center, Dallas, TX, USA

Jeremy Dkhar AMACIP Division, CSIR-Institute of Himalayan Bioresource Technology, Palampur, Himachal Pradesh, India

Neeraj Kumar Dubey Botany Department, Rashtriya Snatkottar Mahavidyalaya, Jamuhai, Jaunpur, UP, India

Elnaz Ghotbi Department of Dermatology, University of Texas Southwestern Medical Center, Dallas, TX, USA

Reena Jain Department of Chemistry, Hindu College, University of Delhi, Delhi, India

Neill Kim Department of Chemistry and Biochemistry, Texas Tech University, Lubbock, TX, USA

Skalzang Lhamo Department of Botany, Central University of Jammu, Samba, Jammu and Kashmir, India

Bendangchuchang Longchar Department of Life Science, Pachhunga University College, Mizoram University, Aizawl, Mizoram, India

Shakti Mehrotra Department of Biotechnology, Institute of Engineering and Technology, Lucknow, Uttar Pradesh, India

Sonal Mishra School of Biotechnology, University of Jammu, Jammu, Jammu and Kashmir, India

Avinash Kaur Nagpal Department of Botanical Environmental Sciences, Guru Nanak Dev University, Amritsar, Punjab, India

Pratap Kumar Pati Department of Biotechnology, Guru Nanak Dev University, Amritsar, Punjab, India

Asosii Paul Department of Botany, Nagaland University, Lumami, Nagaland, India

Pravin Prakash Molecular Biology & Biotechnology Division, Council of Scientific and Industrial Research-National Botanical Research Institute (CSIR-NBRI), Lucknow, Uttar Pradesh, India

Savita Department of Botany, Hindu College, University of Delhi, Delhi, India

Shital K. Sharma Bio-processing and Herbal Division, Mahatma Gandhi Institute for Rural Industrialization, Wardha, Maharashtra, India

Ankita Singh Botany Department, Rashtriya Snatkottar Mahavidyalaya, Jamuhai, Jaunpur, UP, India

Prashant Singh Botany Department, Rashtriya Snatkottar Mahavidyalaya, Jamuhai, Jaunpur, UP, India

Anju Srivastava Department of Chemistry, Hindu College, University of Delhi, Delhi, India

Rakesh Srivastava Molecular Biology & Biotechnology Division, Council of Scientific and Industrial Research-National Botanical Research Institute (CSIR-NBRI), Lucknow, Uttar Pradesh, India

Vikas Srivastava Department of Botany, Central University of Jammu, Samba, Jammu and Kashmir, India

Charles Stewart, Jr Office of Biotechnology, Iowa State University, Ames, IA, USA

Praveen Chandra Verma Molecular Biology & Biotechnology Division, Council of Scientific and Industrial Research-National Botanical Research Institute (CSIR-NBRI), Lucknow, Uttar Pradesh, India

Satyendra Kumar Yadav Zoology Department, Rashtriya Snatkottar Mahavidyalaya, Jamuhai, Jaunpur, UP, India

Part I

Tropane Alkaloids: Diversity, Biosynthesis and Significance

Plant Tropane Alkaloids (TAs): Chemotaxonomic Significance, Biogenesis, and Biotechnological Interventions

1

Savita, Anju Srivastava, Reena Jain, Avinash Kaur Nagpal, and Pratap Kumar Pati

Abstract

Tropane alkaloids (TAs) are valuable secondary metabolites widely distributed in plants under natural conditions. All TAs contain a unique bicyclic tropane ring system. Several tropane alkaloids have medicinal value; e.g., atropine is used to treat certain types of nerve agents, reduces rigidity in Parkinsonism, and can be used as an antidote to poisoning with parasympathomimetic agents and organophosphorus insecticides. Another TA, scopolamine, is used in preventing nausea and vomiting. Tropane alkaloids exhibit a pattern and distribution frequencies that establish them as chemotaxonomic markers in many angiospermic families. This chapter illustrates the chemotaxonomic significance of tropane alkaloids, recent achievements in understanding of tropane alkaloid biosynthesis especially at enzymatic level, and modulation of tropane alkaloids through biotechnological interventions.

Keywords

Tropane alkaloids · Chemotaxonomy · Biosynthesis · Solanaceae · Erythroxylaceae

Savita
Department of Botany, Hindu College, University of Delhi, Delhi, India

A. Srivastava · R. Jain
Department of Chemistry, Hindu College, University of Delhi, Delhi, India

A. K. Nagpal
Department of Botanical and Environmental Sciences, Guru Nanak Dev University, Amritsar, Punjab, India

P. K. Pati (✉)
Department of Biotechnology, Guru Nanak Dev University, Amritsar, Punjab, India

V. Srivastava et al. (eds.), *Tropane Alkaloids*,
https://doi.org/10.1007/978-981-33-4535-5_1

1.1 Introduction

Secondary metabolites are widely distributed in different plant families, and they exhibit wide variation in their chemical structure, distribution, and function (Smith 1976; Pagare et al. 2015). There are several secondary metabolites produced by plants such as alkaloids, pigments, terpenoids, tannins, resins, gums, phenolics, and nonprotein amino acids, and they are utilized for chemotaxonomic classification. Chemotaxonomic studies are helpful to taxonomists, to solve selected taxonomical problems which rely on the chemical similarity of taxa (Hegnauer 1986). Alkaloids are heterocyclic nitrogen-containing basic compounds that contain secondary, tertiary, or quaternary nitrogen atoms in their molecules (Atal 1982). Alkaloids such as tropane alkaloids, indole alkaloids, and isoquinoline alkaloids have also been a useful tool for taxonomic classification of plants (Bentley 1992). Tropane alkaloids are widely distributed in angiospermic families, and more than 200 tropane alkaloids are known in various plant species (Georgiev et al. 2013). All TAs contain a unique bicyclic tropane ring system [(8-azabicyclo[3.2.1]octane nucleus] as a key structural element, and there are three major groups of TAs: hyoscyamine and scopolamine, cocaine, and calystegines (Griffin and Lin 2000). Hyoscyamine and scopolamine are the main active principle from *Solanaceae*. *Solanaceae* is considered as the home of tropane alkaloids, but they have also been identified in other angiospermic families such as *Convolvulaceae*, *Cruciferae*, *Erythroxylaceae*, *Euphorbiaceae*, *Moraceae*, *Proteaceae*, and *Rhizophoraceae* (Griffin and Lin 2000; Kohnen-Johannsen and Kayser 2019). A lot of reports are published about the pharmacological effects of tropane alkaloids, but little is known about the biosynthetic pathways and their regulation. The progress in elucidation of biosynthetic pathways is hampered since long due to nonavailability of good model system. Plant tissue culture and hairy root culture techniques have been proven to be important tools in the elucidation of biosynthetic pathways of secondary metabolites. These techniques have also been widely utilized as an alternative production system for tropane alkaloids. In this chapter, chemotaxonomic significance, recent achievements in tropane alkaloid biogenesis at enzymatic level, modulation of tropane alkaloids through biotechnological interventions, and metabolic engineering are described.

1.2 Tropane Alkaloids in Chemotaxonomy

Chemotaxonomy is the science in which the information about the chemical constituents of plants and the differences in their biochemical features are used for the classification of plants. Classification of plants began as edible and inedible on the basis of their nutritive and organoleptic properties. However, these properties are also the result of their chemical composition. Many a times, only a part(s) of plant possesses these properties due to their specific chemical composition (Griffin and Lin 2000). Chemotaxonomy has become a modern trend for the classification of plants due to recent developments in biotechnology, biochemistry, and phytochemistry which facilitate the study of biological molecules, especially secondary metabolites. Secondary metabolites have special applications in plant taxonomy because the chemical structure of secondary metabolites is often specific and restricted to taxonomically related plants. Quite often, the biosynthetic pathways producing these compounds differ in various plant groups, and this information can be utilized for the chemotaxonomic purpose. The information about the chemical structure of secondary metabolites and their biosynthetic pathways support the existing schemes of botanical classification based on morphology and phylogeny. The natural chemical constituents of the plants or any other life forms can be classified as:

a. *Primary metabolites*: These molecules are involved in vital metabolic pathways and are formed during the growth phase as a result of energy metabolism. Common examples of primary metabolites include alcohol (ethanol), amino acids (L-glutamate and L-lysine), organic acids (aconitic acid and citric acid), etc.
b. *Secondary metabolites*: Compounds which are derived from primary metabolites and often perform non-vital functions are known as secondary metabolites, such as alkaloids, glucosinolates, nonprotein amino acids, gums, resins, phenolic compounds, terpenes, etc.

1.3 Chemotaxonomy of Plant Families Containing Tropane Alkaloids

Tropane alkaloids share a similar basic structure, but they have different biochemical and pharmacological properties. Tropane alkaloids exhibit a pattern and distribution frequency that establishes them as chemotaxonomic markers in many angiospermic families.

1.3.1 Family: *Solanaceae*

Solanaceae is the family which is known as the home of tropane alkaloids because of their abundance in the family. This family is usually divided into different subfamilies and tribes, characterized by chemical differences in their composition. In 1987, Tétényi classified the *Solanaceae* family into four subfamilies (*Anthocercoideae*, *Cestroideae*, *Solanoideae*, and *Atropoideae*) based on the occurrence of alkaloids and steroids (Tétényi 1987).

Currently two proposed classifications of Solanaceae are accepted. The first, which is based on morphological and chemical criteria, classifies Solanaceae into six subfamilies: (1) *Solanoideae*, (2) *Cestroideae*, (3) *Juanulloideae*, (4) *Salpiglossoideae*, (5) *Schizanthoideae*, and (6) *Anthocercidoideae* (Hunziker 2001).

The second and most accepted proposal divides *Solanaceae* into seven subfamilies: (1) *Solanoideae*, (2) *Cestroideae*, (3) *Nicotianoideae*, (4) *Petunioideae*,

(5) *Schizanthoideae*, (6) *Goetzeoideae*, and (7) *Schwenckioideae* (Olmstead et al. 2008).

In an earlier classification by Hunziker (1979), *Solanaceae* was subdivided into two subfamilies—*Solanoideae* and *Cestroideae*—which were further subdivided into tribes on the basis of the distribution of tropane alkaloids within the family (Hunziker 1979).

a. Subfamily: *Solanoideae*

This subfamily is divided into several well-established tribes based on distribution of tropane alkaloids:

1. *Tribe Datureae*: This tribe consists of two genera—*Datura* and *Brugmansia*. The aerial parts of *Datura* species contain hyoscyamine, scopolamine, littorine, and the base 6β-(2-methylbutanoyloxy) tropan-3α-ol, while the roots yield valtropine. In *Brugmansia*, the aerial parts contain scopolamine and/or hyoscyamine as principal alkaloids, with smaller amounts of derivatives of these bases. The roots contain in addition a large number of esters formed from dihydroxytropane and teloidine (Griffin and Lin 2000).
2. *Tribe Hyoscyameae*: *Scopolia* spp. and *Hyoscyamus* spp. contain hyoscyamine as major alkaloid along with scopolamine derivatives.
3. *Tribe Solandreae*: The genus *Solandra* contains atropine, littorine, hyoscyamine, and their derivatives as principal alkaloids.
4. *Tribe Solaneae*: *Atropa belladonna* L. contains hyoscyamine, scopolamine, and apoatropine as the principal alkaloids (Evans 1979). The minor genera *Latua* and *Acristus* contain hyoscyamine and tigloidine (Evans 1979). *Physalis alkekengi* contains tigloidine, 3α-tigloyloxytropane, cuscohygrine, and phygrine (hygrine dimer) in its roots (Basey and Woolley 1973). Instead of hyoscyamine and scopolamine, *Solanum* sp. contain calystegine A3 along with calystegine B2 in the leaves (Evans 1979). *Mandragora* sp. contain hyoscyamine, scopolamine, cuscohygrine, apoatropine, 3α-tigloyloxytropane, and 3,6-ditigloyloxytropane (Jackson and Berry 1973). Roots of *Salpichroa origanifolia* contain small quantities of cuscohygrine, pseudotropine, tropine, and possibly hyoscyamine (Evans et al. 1972). *Withania somnifera* likewise contains cuscohygrine, 3-α-tigloyloxytropane, tropine, and pseudotropine (Leary et al. 1963). The main base from the roots of *Cyphomandra betaceae* was N,N-′-bis-(4-dimethylaminobutyl) hexamide, tropinone, cuscohygrine, hyoscyamine, tigloidine, tropine, and pseudotropine.
5. *Incertisadis*: Genera with doubtful taxonomic positions such as *Duckeodendron* Kuhlmannb, *Parabouchetia* Baillon, and *Pauia* Deb. and Dutta.

b. Subfamily: *Cestroideae* (*Browallioideae*)
   1. *Tribe Anthocercideae*: The tribe *Anthocercideae* includes seven genera: *Anthocercis*, *Anthotroche*, *Crenidium*, *Cyphanthera*, *Duboisia*,

Grammosolen, and *Symonanthus*. These genera differ from each other on the basis of their tropane alkaloid contents.

Anthocercis littorea mainly contains littorine and meteloidine and tropane derivatives such as mono- and ditigloyl esters of teloidine; *A. viscosa* R. Br. and *A. fasciculata* F. Muell. contain hyoscyamine as predominant alkaloids (El-Imam and Evans 1984; Cannon et al. 1969).

Anthotroche myoporoides, *A. pannosa*, and *A. walcottii* contain hyoscyamine, norhyoscyamine, apoatropine, and scopolamine as predominant alkaloids (Evans and Ramsey 1981).

Crenidium spinescens Haegi contains hyoscyamine as the predominant alkaloid, but anabasine with ursolic acid is also present (El-Imam and Evans 1984).

Cyphanthera anthocercidea Haegi (*Anthocercis frondosa*) contains hyoscyamine, scopolamine, and nicotine (more abundant). *Cyphanthera albicans* Miers (*Anthocercis albicans*) contains butyryl esters of tropine and 6-β-hydroxytropine as predominant alkaloids (El-Imam and Evans 1984).

Duboisia myoporoides contains scopolamine; *D. leichhardtii* F. Muell contains hyoscyamine, scopolamine, and calystegine; *D. arenitensis* contains hyoscyamine and scopolamine as the most important tropane alkaloids (Evans and Ramsey 1981).

Grammosolen spp. is most closely related chemically to *Cyphanthera* and *Anthotroche* (Evans and Ramsey 1981; Evans and Ramsey 1983).

Symonanthus aromaticus Haegi contains mono- and ditigloyl esters. Scopolamine and its derivative aposcopolamine are found as the main alkaloids in both aerial parts and roots (Evans and Ramsey 1983).

2. *Tribe Nicandreae*: *Nicandra* sp. does not contain esters of tropic acid. However, tropine was isolated from the roots (Parr 1992).
3. *Tribe Salpiglossideae*: *Schizanthus pinnatus* Ruiz contains tropane derivatives such as schizanthines A and B which are 6β-senecioyloxytropan-3β-ol esters of dibasic mesaconic acid. *Schizanthus litoralis* contains hydroxytropane esters, hygrolines, and the tropane diester of itaconic acid. *S. grahamii* contains schizanthines C, D, and E and trimeric tropane alkaloids (San Martin et al. 1987).

1.3.2 Family: *Erythroxylaceae*

Erythroxylaceae is also known as the coca family and is a group of flowering trees and shrubs consisting of 4 genera and 271 species. The genus *Erythroxylum* is the best known including the species *E. coca*, the source of the drug cocaine. The comparative phytochemistry of Erythroxylaceae was reviewed (Evans 1981). *Erythroxylum* consists of 19 sections. The alkaloid content in six species of *Erythroxylum* has been reported (Schulz 1907).

Of the six species, *E. mamacoca* contains 3β-benzoyloxytropane together with nortropacocaine (3β-benzoyloxynortropane) (El-Imam et al. 1985). *E. coca*, *E. novogranatense*, and its variety contain dihydrocuscohygrine and cuscohygrine. *E. cataractarum* contains cuscohygrine as the major alkaloid. A new alkaloid tropacocaine was found in *E. ulei* (section Leptogramme) and in *E. mamacoca* and *E. argentinum* (section Archerythroxylon) but not in the sections Marcrocalyx and Rhabdophyllum. Cocaine and the *cis*- and *trans*-isomers of cinnamoyl cocaine were reported in *E. coca* and *E. novogranatense* var. truxillense (Rivier 1981; Zuanazzi et al. 2001).

In *Solanaceae*, the production of atropine and scopolamine is catalyzed by methylecgonone reductase (short chain dehydrogenase/reductase family). In *Solanaceae*, this enzyme is found in root tissues, while it is not found in *Erythroxylaceae* (Jirschitzka et al. 2012). This evidence proves that during the evolution of angiosperms, the ability to produce tropane alkaloids has arisen more than once.

1.3.3 Family: *Proteaceae*

Proteaceae is a fairly large family, which contains around 75 genera with 1700 species divided into 5 subfamilies and 14 tribes (Johnson and Briggs 2008). The subfamilies of this family are as follows: (1) *Bellendenoideae* (*Bellandena*); (2) *Persoonioideae* contains two tribes: *Placospermeae* and *Persoonieae*; (3) *Symphionematoideae*; (4) *Proteoideae* contains four tribes: (a) *Conospermeae*, (b). *Petrophileae*, (c). *Proteeae*, and (d) Leucadendreae; and (5) *Grevilleoideae* contains four tribes: (a) *Roupaleae*, (b) *Banksieae*, (c) *Embothrieae*, and (d) *Macadamieae*.

Bellendine, an unusual tropane alkaloid, was first isolated from *Bellendena montana* from the subfamily *Bellendenoideae* of Proteaceae family (Bick et al. 1971). Other similar alkaloids including darlingine as the major alkaloid were isolated from *Darlingia darlingiana* from the subfamily *Grevilleoideae* (Bick et al. 1979a). *D. ferruginea* contains darlingine, ferruginine, ferrugine, and 3-α-benzoyloxy-2α-hydroxybenzyltropine (Bick et al. 1979b). *Knightia strobilina* (subfamily *Grevilleoideae*) contains strobiline as the major alkaloid, but it also contains 3α-cinnamoyloxytropan-6β-ol and 3α-acetoxy-2α-acetoxybenzyltropane (acetylknightionl), 3α-acetoxy-2α-benzyltropan-6β-ol (knightoline), 6-β-benzoyloxytropan-3α-ol, 2α-hydroxybenzyl-3α-acetoxytropane (knightinol), and dihydrostrobiline (Lounasmaa et al. 1980). *Agastachys odorata* (subfamily *Persoonioideae*) contains 6β-acetoxy-3α-tigloyloxytropane and 3-α-(p-hydroxybenzoyloxy) trop-6-ene (Bick et al. 1979c).

1.3.4 Family: *Euphorbiaceae*

Euphorbiaceae is the fifth largest family of flowering plants containing three subfamilies (*Acalyphoideae*, *Crotonoideae*, and *Euphorbioideae*), 37 tribes, 300 genera, and 7500 species.

Pyrrolidine and tropane alkaloids have been reported from *Croton gabouga* S. Moore. Similar tropane alkaloids were also isolated from *Margaritaria discoidea* (Baill.) Webster (syn. *Phyllanthus discoideus* Mull. Arg.) which was earlier a member of *Euphorbiaceae* family but now placed in Phyllanthaceae (Seigler 1977).

1.3.5 Family: *Elaeocarpaceae*

Elaeocarpaceae contains 12 genera and approximately 615 species of trees and shrubs. Only *Peripentadenia mearsii* contains the tropane alkaloids such as tropacocaine (3β-benzoyloxytropane), 2α-benzoyloxy-3β-hydroxytropane, and 3-α-acetoxy-6β-hydroxytropane (Johns et al. 1971).

1.3.6 Family: *Rhizophoraceae*

Rhizophoraceae contains 15 genera and around 147 species and is better known for mangrove trees of *Rhizophora* species. *Rhizophoraceae* contains the dithiolane esters (Jirschitzka et al. 2013). Tropane alkaloids obtained from members of *Rhizophoraceae* are as follows:

Bruguiera sexangula (a tropical mangrove) contains brugine (tropine 1,2-dithiolane-3-carboxylate); *B. exaristata* contains tropine esters of acetic, benzoic, n-butyric, isobutyric, propionic, and isovaleric acids (Loder and Russel 1966; Loder and Russel 1969; Gnecco et al. 1983). *Crossostylis* spp. (*C. biflora*, *C. multiflora*, and *C. sebertii*) also contain brugine along with tropine, 3α-benzoyloxytropane, tropine cinnamate, and ferulate (Gnecco et al. 1983).

1.3.7 Family: *Convolvulaceae*

Convolvulaceae is commonly known as the bindweed or morning glory family and consists of 60 genera and more than 1650 species (Irshad et al. 2015). The characteristic tropane alkaloids found in the *Convolvulaceae* include convolvine, convolidine, confoline, and convolamine N-oxide (Sharova et al. 1980; Aripova 1985). Roots of *Convolvulus arvensis* and *Calystegia sepium* contain calystegines B1, B2, and A3 (Sharova et al. 1980; Aripova 1985). *Evolvulus sericeus* contains convoline, convolamine, and convolidine (Conselo and Alejandra 1972). *Erycibe obtusifolia* contains baotongteng A, (2β-hydroxy-6β-acetoxynortropane) (Yao et al. 1981) and baotongteng B (Seigler 1977). *E. elliptilimba* and *E. hainanensis* contain erycibelline (2β,7β-dihydroxynortropane) (Chen et al. 1986; Lu et al. 1986; Wang et al. 1990).

1.3.8 Family: *Cruciferae*

Cruciferae (*Brassicaceae*) family consists of 338 genera and 3709 species (Al-Shehbaz et al. 2006). Cochlearine, tropine, and pseudotropine were found in *Cochlearia arctica* and other ten species of *Cochlearia* (Platonova and Kusovkov 1963; De Simone and De Feo Margarucci 2008).

1.3.9 Family: *Moraceae*

Moraceae commonly known as mulberry family consists of 39 genera and 1100 species (Zerega et al. 2010). *Morus alba* L. is the best known member of *Moraceae* which contains tropane alkaloids. Root bark of *M. alba* contains penthydroxylated nortropane calystegine C1, while fruits contain calystegines A8 and A9.

1.4 Tropane Alkaloids as Chemotaxonomic Evidence

Chemotaxonomic classification based on tropane alkaloids supports the modern botanical classification which is based on molecular phylogeny (Olmstead et al. 2008). In recent phylogenetic classification (APG-III), *Solanaceae* and *Convolvulaceae* are placed in the order of Solanales. Both the families contain calystegines B1, B2, and A3 in common which is evidence in support of the phylogenetic classification (Sharova et al. 1980; Aripova 1985). Now, in fact, *Solanaceae* and *Convolvulaceae* are considered "sisters" because of their morphology and their plastid DNA (Pigatto et al. 2015; Judd et al. 1999; Downie and Palmer 1992; Olmstead et al. 1992; Olmstead et al. 1993; Taktajan 1997; Thorne 1992; Cronquist 1981). The *Nicandra* genus is related to *Solanum* and *Physalis*, which are known for the production of calystegines (CA) and are located on more derived branches of the Olmstead phylogeny (D'Arcy 1979). In phylogenetic classification, the genus *Latua* was moved to the subfamily *Solanoideae* (*Atropina clade*), which is consistent with the chemical data reported for this genus (Olmstead et al. 2008). In an earlier classification, *Latua* was moved to the *Jaboroseae* tribe of the *Solanoideae* subfamily, due to the presence of tropane alkaloids, among other characteristics (Tétényi 1987). In the Olmstead phylogeny, the genus *Jaborosa* lies close to *Latua*, supporting these previous findings. *Euphorbiaceae* and *Rhizophoraceae* are placed in the order Malpighiales. Both the families share in common the tropine esters of acetic and benzoic acids (Bick et al. 1971). Moreover, in the modern phylogenetic classification, *Brassicaceae* (Brassicales), *Moraceae* (Rosales), and *Malpighiales* members are placed in the same group *Roside* (Judd et al. 1999). The above findings corroborate with the recent phylogenetic system of classification (APG-III, 2016).

Based on the above data, it can be clearly stated that biosynthesis of similar secondary metabolites can be related to a common phylogenetic origin.

1.5 Biogenesis of Tropane Alkaloids

Edward Leete started the pioneering studies on the biosynthesis of tropane alkaloids using whole plant and isotope labels (Leete 1990). In spite of intensive studies, the role of specific enzymes in the biosynthesis of tropane alkaloids is not totally understood. Reciprocal grafting between tropane alkaloid-producing and tropane alkaloid-nonproducing plants has proven that the roots are the site of alkaloid biosynthesis, from where these alkaloids are transported to the other parts of plants through the xylem. The ratio of these alkaloids in different plant parts may vary according to different species (Waller and Nowachi 1978; Hashimoto et al. 1991; Samuelsson 1999). The biosynthesis of tropane alkaloids such as hyoscyamine and scopolamine has been explained at the enzyme level (Fig. 1.1). The bicyclic tropane ring is derived from L-ornithine and/or L-arginine via tropinone, whereas the tropic acid moiety is synthesized from phenylalanine. Putrescine is a common precursor of both polyamines (spermidine and spermine) and tropane/pyridine alkaloids (Robins et al. 1990; Hashimoto and Yamada 1987). Ornithine and arginine can be converted to putrescine by the activity of enzymes ornithine decarboxylase (ODC) and arginine decarboxylase (ADC), respectively. Ornithine decarboxylase (ODC) isolated from

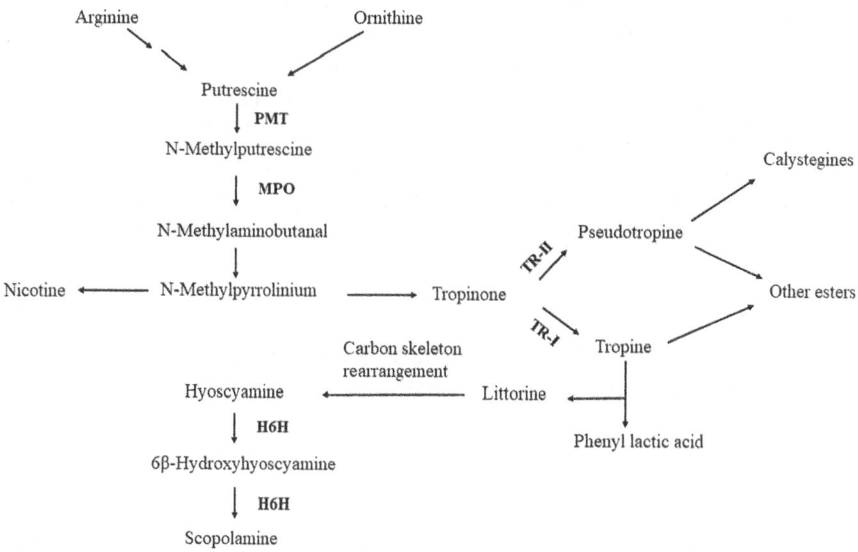

Fig. 1.1 The outline of biosynthetic pathways of tropane alkaloids. Abbreviations: *PMT* putrescine N-methyltransferase, *MPO* methylputrescine oxidase, *TRI* tropinone reductase I, *TRII* tropinone reductase II, *H6H* hyoscyamine-6β-hydroxylase

Hyoscyamus albus results in the conversion of ornithine to putrescine. Arginine decarboxylase (ADC) results in the conversion of arginine into agmatine leading via N-carbamoyl putrescine to putrescine (Hibi et al. 1992). Putrescine is converted to N-methylputrescine with the activity of the enzyme putrescine N-methyltransferase (PMT). This is the first crucial step in the biosynthesis of tropane alkaloids and nicotine, because the pyrrolidine ring of nicotine and the tropane ring moiety of the tropane alkaloids are derived from N-methylputrescine synthesis (Matsuda et al. 1991). N-methylputrescine oxidase (MPO) isolated from *H. niger* is a diamine oxidase enzyme which catalyzes the formation of N-methyl-Δ^1-pyrrolinium cation from N-methylputrescine via 4-methylaminobutanal. It is evident from the experiments that N-methyl-Δ^1-pyrrolinium salt is a precursor of the tropane nucleus (Leete 1990; Feth et al. 1985). N-methyl-Δ^1-pyrrolinium gets converted spontaneously into tropinone without catalyzation by any enzyme. Tropinone reductases (TRs) are the enzymes responsible for the reduction of the keto group in the tropinone (Dräger 2006). In *Solanaceae*, two separate tropinone reductases are found: TRI and TRII. TRI converts the tropinone into tropine (3α-tropanol). Tropine is the precursor for the formation of various esterified tropane alkaloids. On the other hand, TRII converts tropinone to pseudotropine (3β-tropanol), which is the precursor for the formation of various nonesterified tropane alkaloids called calystegines. Hyoscyamine and scopolamine are the common tropane alkaloids. The biosynthesis of TAs involves the activation of phenyllactic acid, likely through the formation of phenyllactoyl-CoA (Robins et al. 1994). Phenyllactoyl-CoA could transfer its acyl group to tropine to form littorine. The last step in the tropane alkaloid biosynthesis is the carbon skeleton rearrangement of littorine to form hyoscyamine, which is processed by a mutase (Chesters et al. 1995). Hyoscyamine is converted to scopolamine, which is the 6,7-β-epoxide of hyoscyamine. This reaction is catalyzed by hyoscyamine-6-β-hydroxylase (H6H) which is a bifunctional enzyme. The H6H is a 2-oxoglutarate-dependent dioxygenase that requires alkaloid substrate, 2-oxoglutarate, Fe^{2+}, molecular oxygen, and ascorbate for catalysis. H6H catalyzes the hydroxylation of hyoscyamine to 6-β-hydroxyhyoscyamine, as well as the epoxidation of 6-β-hydroxyhyoscyamine to scopolamine (Hashimoto and Yamada 1987).

1.6 Modulation of Tropane Alkaloids Through Biotechnological Interventions

Various biotechnological approaches are being used to modulate the production of tropane alkaloids (Hashimoto and Yamada 1987; Palazón et al. 2008; Oksman-Caldentey and Arroo 2000). Hairy root (HR) cultures have been extensively used for the enhanced production of secondary metabolites such as tropane alkaloids (Harfi et al. 2018; Ajungla et al. 2009; Palazón et al. 2008; Srivastava et al. 2018; Srivastava et al. 2020). Hairy root culture is induced by genetic transformation with *Agrobacterium rhizogenes* which has been successfully used in the in vitro production of hyoscyamine and scopolamine (Zhang et al. 2007; Ghorbanpour et al.

2013; Zhao et al. 2020). There are three classical strategies to increase secondary metabolite production through plant cell and tissue cultures:

1. The screening and selection of lines with secondary metabolite production (Giri and Zaheer 2016)
2. Optimization of growth and production media (Giri and Zaheer 2016)
3. Elicitation of secondary metabolite production by abiotic or biotic elicitors and metabolic engineering (Giri and Zaheer 2016)

Elicitation is a strategy enhancing the production of tropane alkaloids and other secondary metabolites in the adventitious and hairy root cultures of many plants. Since most of the secondary metabolites are produced in response to exposure with various biotic and abiotic stresses, they provide protection to plants and help them to adapt or manage in adverse environmental conditions. Accumulation of secondary metabolites is observed in plants on exposure to abiotic stresses or with interaction with microbes or their products. Elicitors are the substances which are used in small quantity in living systems to induce or enhance the production of secondary metabolites which play important role in the adaptations of plants to different types of stresses (Giri and Zaheer 2016; Biondi et al. 2002; Ryad et al. 2010; Oksman-Caldentey and Arroo 2000). The common elicitor molecules include methyl jasmonate (MeJa), salicylic acid (SA), and acetyl salicylic acid (ASA), which are frequently used in elicitation experiments to improve alkaloid production rate to favor their commercial exploitation (Giri and Zaheer 2016). PMT (putrescine N-methyltransferase) is the first critical enzyme involved in the biosynthetic pathways of tropane alkaloids which is encoded by the *pmt* gene. The expression of *pmt* gene is controlled by light, hormones, biotic and abiotic stresses, and elicitors like jasmonates. There are several reports which indicate the potential role of methyl jasmonate (MeJa) in the elicitation of tropane alkaloid production (Zhang et al. 2007; Kang et al. 2004; Harfi et al. 2018; Oksman-Caldentey and Arroo 2000). It was reported that the *pmt* gene may have jasmonate- and elicitor-responsive element (JERE) in its promoter region which makes it responsive toward MeJa treatment and improves the accumulation of tropane alkaloids in root cultures of *Nicotiana sylvestris (*Shoji et al. 2000*)*, *Hyoscyamus niger (*Zhang et al. 2007*)*, *H. muticus (*Biondi et al. 2002*)*, and *Datura stramonium (*Ryad et al. 2010*)*. On the contrary, the treatment of *A. belladonna* roots (Hibi et al. 1992) and callus cultures of *H. muticus (*Biondi et al. 2002*)* with MeJA did not affect secondary metabolism. Kang et al. reported the overexpression of *PMT* and *H6H* genes by elicitation with methyl jasmonate and salicylic acid in adventitious root cultures of *Scopolia parviflora* (Kang et al. 2004). Increased production of hyoscyamine was reported in *Datura* spp. by the treatment of elicitors such as SA and ASA in the hairy root cultures (Harfi et al. 2018). Ajungla et al. (2009) studied the effect of biotic and abiotic elicitors on the production of hyoscyamine and scopolamine in root cultures of *Datura metel*. They reported that tropane alkaloid production in hairy roots (HRs) increased by 3.13-fold with the SA elicitor treatment while the improvement factor of tropane alkaloids in HRs is function of elicitor, with the following order: methyl jasmonate, yeast extract, and oligogalacturonides (Ajungla et al. 2009). They also studied the effect of fungal elicitors such as *Aspergillus niger*, *Alternaria* spp., and *Fusarium monoliforme* on the production of hyoscyamine and scopolamine in *Datura metel*, and it was found that the accumulation of these alkaloids increased with the increasing concentration of homogenate of fungal elicitors (Ajungla et al. 2009). Pitta-alvarez et al. reported enhanced accumulation of hyoscyamine and scopolamine in root cultures of *Brugmansia candida* by elicitation with salicylic acid, silver nitrate, yeast extract, $CaCl_2$, and $CdCl_2$ (Pitta-Alvarez et al. 2000). Ballica et al. reported the increased production of tropane alkaloids by the addition of cell wall fragments of *Phytophthora megasperma* (Pmg) and by increasing carbon/nitrogen ratio up to about 100 in suspension cultures of *Datura stramonium* (Ballica et al. 1993). Ghourbanpour et al. reported that rhizobacterial strains such as *Pseudomonas putida* (PP) and *Pseudomonas fluorescens* (PF) significantly increased the accumulation of hyoscyamine and scopolamine in *Hyoscyamus niger* under water deficit stress (Ghorbanpour et al. 2013). Increased calcium content in Murashige and Skoog's (MS) medium enhanced the accumulation of scopolamine while decreasing the hyoscyamine accumulation in *Hyoscyamus niger* (Pudersell et al. 2012). Zhao et al. reported that transgenic *Atropa belladonna* plants overexpressing *AbODC* (*ODC* gene from *A. belladonna*) showed a significantly higher production of hyoscyamine and anisodamine compared with control plants (Zhao et al. 2020).

1.7 Conclusion and Future Prospectus

Chemical and biological resemblances among angiospermic families containing tropane alkaloids support the modern phylogenetic classification (APG-III). Hyoscyamine, scopolamine, and calystegines have been established as molecular markers for the classification of families which contain tropane alkaloids. They have proved helpful to taxonomists to solve the taxonomical problems. The above facts support that biosynthesis of similar secondary metabolites (tropane alkaloids) can be related to a common phylogenetic origin.

Hairy root culture and metabolic engineering techniques have proven their potential to improve the pharmacological properties of medicinal plants quantitatively and qualitatively. However, these methods require the complete knowledge of steps of the biosynthetic pathways and the respective genes involved in the regulation of expression of desired medicinal products. Another problem is the presence of multiple rate-limiting steps which make it difficult to predict the result of overexpression of a single gene or multiple genes involved in the biosynthetic pathways. Instead of focusing on a single gene, it becomes necessary to consider the homeotic genes controlling the expression of numerous genes involved in the biosynthetic pathways which can lead to the regulation of whole pathway and interconnecting cellular pathways. As we know, the secondary metabolite production is strictly under regulation of transcription factors which can increase the expression of a series of enzymes involved in a metabolic pathway. Use of transcription factors can avoid the time-consuming steps of finding the role of individual enzyme in poorly known biosynthetic pathways (Gantet and Memelink 2002). Recently, there have been reports indicating the potential use of elicitor molecules such as MeJa, SA, fungal elicitors, etc. in the elicitation of expression of many genes involved in tropane alkaloid biosynthesis. In addition to elicitor molecules, antisense genes can also be used to block the competitive pathways to increase the flux of the desired secondary metabolites (Chintapakorn and Hamill 2003).

References

Ajungla L, Patil PP, Barmukh RB, Nikam TD (2009) Influence of biotic and abiotic elicitors on accumulation of hyoscyamine andscopolamine in root cultures of *Datura metel* L. Indian J Biotechnol 8:317–322

Al-Shehbaz IA, Beilstein MA, Kellogg EA (2006) Systematics and phylogeny of the Brassicaceae (Cruciferae): an overview. Plant Syst Evol 259:89–120

Aripova SF (1985) N-Oxide of convolamine from *Convolvulus krauseanus*. Khimiya Prirodnykh Soedinenii 2:275

Atal C (1982) Cultivation and utilization of aromatic plants, 1st edn. Council of Scientific and Industrial Research, New Delhi, pp 15–21

Ballica R, Ryu DDY, Kado CI (1993) Tropane alkaloid production in *Datura stramonium* suspension cultures: elicitor and precursor effects. Biotechnol Bioeng 41:1075–1081

Basey K, Woolley JG (1973) Alkaloids of *Physalis alkekengi*. Phytochemistry 12:2557–2559

Bentley KW (1992) β-Phenylethylamines and the isoquinoline alkaloids. Nat Prod Rep 9:365–391

Bick IRC, Bremer JB, Gillard JW (1971) Methyl(p-hydroxybenzoyl)acetate and an alkaloid, bellendine, from *Bellendena montana*. Phytochemistry 10(2):475–477

Bick IRC, Gillard JW, Huck-Meng L (1979a) Alkaloids of *Bellendena montana*. Aust J Chem 32 (8):1827–1840

Bick IRC, Gillard JW, Huck-Meng L (1979b) Aust J Chem 32:2537

Bick IRC, Gillard JW, Huck-Meng L, Preston W (1979c) Aust J Chem 32:2071

Biondi S, Scaramagli S, Oksman-Caldentey KM (2002) Secondary metabolism in roots and callus cultures of *Hyoscyamus muticus* L.: the relationship between morphological organization and the response to methyl jasmonate. Plant Sci 163:563–569

Cannon JR, Joshi KR, Meehan GV, Williams J (1969) The tropane alkaloids from three western Australian *Anthocerus species*. Aust J Chem 22:221

Chen Z, Xu P, Yao T (1986) YaoxueXuebao, 21, 513 (Ch.), via C.A., 106, 153016

Chesters NJCE, O'Hagan D, Robins RJ (1995) The biosynthesis of tropic acid: the (R) -D-phenyllactoyl moiety is processed by the mutase involved in hyoscyamine biosynthesis in *Datura stramonium*. J Chem Soc Chem Commun 1995:127–128

Chintapakorn Y, Hamill JD (2003) Antisense-mediated downregulation of putrescine *N*-methyltransferase activity in transgenic *Nicotiana tabacum* L. can lead to elevated levels of anatabine at the expense of nicotine. Plant Mol Biol 53:87–105

Conselo FL, Alejandra SG (1972) Rev. Colomb. Cienc. Quim,-Farm., 2, 27 (Span.), via C.A., 79, 15798

Cronquist A (1981) An integrated system of classification of flowering plants. Columbia University Press, New York

D'Arcy WG (1979) The classification of the Solanaceae. In: Hawkes JG, Lester RN, Skelding AD (eds) The biology and taxonomy of the Solanaceae. Academic, London, pp 3–48

De Simone L, De Feo Margarucci V (2008) Tropane alkaloids: an overview. Pharmacol Online 1:70–89

Downie SR, Palmer JD (1992) Restriction site mapping of chloroplast DNA inverted repeat: a molecular phylogeny of the Asteride. Ann Mo Bot Gard 79:266–238

Dräger B (2006) Tropinonereductases, enzymes at the branch point of tropane alkaloid metabolism. Phytochemistry 67:327–337

El-Imam YMA, Evans WC (1984) Tropane alkaloids of species of *Anthocercis, Cyphanthera and Crenidium*. Planta Med 50(1):86–87. https://doi.org/10.1055/s-2007-969628

El-Imam YMA, Evans WC, Plowman T (1985) Alkaloids of some South American Erythroxylum species. Phytochemistry 24:2285–2289

Evans WC (1979) Tropane alkaloids of the Solanaceae. In: Hawkes JG, Lester RN, Skelding AD (eds) The biology and taxonomy of the Solanaceae, Linnean Soc Symp Series No. 7. Academic, New York, p 241

Evans WC (1981) The comparative phytochemistry of the genus Erythroxylon. J Ethnopharmacol 3:265–277

Evans WC, Ramsey KP (1981) Tropane alkaloids from *Anthocercis* and *Anthotroche*. Phytochemistry 20(3):497–499

Evans WC, Ramsey KP (1983) Alkaloids of the solanaceae tribe anthocercideae. Phytochemistry 22(10):2219–2225

Evans WC, Ghani A, Woolley VA (1972) Distribution of littorine and other alkaloids in the roots of *Datura species*. Phytochemistry 11:469–470

Feth F, Arfmann H-A, Wray V (1985) Determination of putrescine N-methyltransferase by high performance liquid chromatography. Phytochemistry 24:921–923

Gantet P, Memelink J (2002) Transcription factors: tools to engineer the production of pharmacologically active plant metabolites. Trends Pharmacol Sci 23:563–569

Georgiev V, Marchev A, Berkov S, Pavlov A (2013) Plant in vitro systems as sources of tropane alkaloids. Nat Prod 2013:173–211

Ghorbanpour M, Hatami M, Khavazi K (2013) Role of plant growth promoting rhizobacteria on antioxidant enzyme activities and tropane alkaloid production of *Hyoscyamus niger* under water deficit stress. Turk J Biol 37:350–360

Giri C, Zaheer M (2016) Chemical elicitors versus secondary metabolite production in vitro using plant cell, tissue and organ cultures: recent trends and a sky eye view appraisal. Plant Cell Tissue Org Cult 126:1–18. https://doi.org/10.1007/s11240-016-0985-6

Gnecco MDH, Pusset M, Pusset J, Husson HP (1983) Tropane alkaloids from *Crossostylis sp*. J Nat Prod 46(3):398–400

Griffin WJ, Lin GD (2000) Chemotaxonomy and geographical distribution of tropane alkaloids. Phytochemistry 53:623–637

Harfi B, Khelifi L, Khelifi-Slaoui M, Assaf-Ducrocq C, Gontier E (2018) Tropane alkaloids GC/MS analysis and low dose elicitors' effects on hyoscyamine biosynthetic pathway in hairy roots of Algerian *Datura* species. Nature 8:17951. https://doi.org/10.1038/s41598-018-36625-4

Hashimoto T, Yamada Y (1987) Purification and characterization of hyoscyamine 6 beta-hydroxylase from root cultures of *Hyoscyamus niger* L. Hydroxylase and epoxidase activities in the enzyme preparation. Eur J Biochem 164:277–285

Hashimoto T, Hayashi A, Amana Y et al (1991) Hyoscyamine 6j3-hydroxylase, an enzyme involved intropane alkaloid biosynthesis, is localized at the pericycle of the root. J Biol Chem 266:4648–4653

Hegnauer R (1986) Phytochemistry and plant taxonomy-an essay on the chemotaxonomy of higher plants. Phytochemistry 25:1519–1535

Hibi N, Fujita T, Hatano M, Hashimoto T, Yamada Y (1992) Putrescine N-methyltransferase in cultures roots of *Hyoscyamus albus*. Plant Physiol 100:826–835

Hunziker AT (1979) The classification of the Solanaceae. In: Hawkes JG, Lester RN, Skelding AD (eds) The biology and taxonomy of the Solanaceae, Linnean Soc. Symp. Series No. 7. Academic, New York, p 49

Hunziker AT (2001) Genera Solanacearum: the genera of Solanaceae illustrated, arranged according to a new system. Gantner, Ruggell, p 500

Irshad S, Pragyadeep S, Rawat AKS, Misra PK, Khatoon S (2015) Comparative pharmacognostical studies of two medicinally important Indian Evolvus species. Indian J Tradit Knowl 14 (4):564–570

Jackson BP, Berry MI (1973) Hydroxytropane tiglates in the roots of *Mandragora species*. Phytochemistry 12:1165–1166

Jirschitzka J, Schmidt GW, Reichelt M, Schneider B, Gershenzon J, D'Auria JC (2012) Plant tropane alkaloid biosynthesis evolved independently in the Solanaceae and Erythroxylaceae. Proc Natl Acad Sci U S A 109(26):10304–10309

Jirschitzka J, Dolke F, D'Auria JC (2013) Chapter two - increasing the pace of new discoveries in tropane alkaloid biosynthesis. Adv Bot Res 68:39–72

Johns SR, Lamberton JA, Sioumis AA (1971) New tropane alkaloids, (+)-(3R, 6R)-3α-acetoxy-6β-hydroxytropane and (+)-2α-benzoyloxy-3β-hydroxynortropane, from Peripentadeniamearsii (Euphorbiaceae). Aust J Chem 24(11):2399–2403

Johnson LAS, Briggs BG (2008) On the Proteaceae - the Evolution and classification of a southern family. Bot J Linn Soc 70(2):83–182

Judd WJ, Campbell CS, Kellogg EA, Stevens PF (1999) Plant systematics. A phylogenetic approach. Sinauer Associates, Inc. Publishers, Sunderland

Kang SM, Jung HY, Kang YM, Yun DJ (2004) Effects of methyl jasmonate and salicylic acid on the production of tropane alkaloids and the expression of PMT and H6H in adventitious root cultures of *Scopolia parviflora*. Plant Sci 166:745–751

Kohnen-Johannsen KL, Kayser O (2019) Tropane alkaloids: chemistry, pharmacology, biosynthesis and production. Molecules 24:796. https://doi.org/10.3390/molecules24040796

Leary JD, Khanna KL, Schwarting AE (1963) Occurrence of cuscohygrine and 3a-tigloyloxytropane in Withania somnifera. Lloydia 26:44–48

Leete E (1990) Recent developments in the biosynthesis of the tropane alkaloids. Planta Med 56:339–352

Loder JW, Russel GB (1966) Tropine 1,2-dithiolane-3-carboxylate, a new alkaloid from *Bruguiera sexangula*. Tetrahedron Lett 51:6327–6329

Loder JW, Russel GB (1969) The alkaloids of *Bruguiera sexangula* and *Bruguiera exaristata* (Rhizophoraceae). Aust J Chem 22:1271

Lounasmaa M, Pusset J, Sevenet T (1980) The New Caledonian plants. Part 57. 2-Benzyl pseudotropanes and dihydropyranotropanes, alkaloids of *Knightia strobilina*. Phytochemistry 19(5):953–955

Lu Y, Yao T, Chen Z (1986) Studies on the constituents of *Erycibe elliptilimba*. Yao xuexuebao 21 (11):829–835

Matsuda J, Okabe S, Hashimoto T, Yamada Y (1991) Molecular cloning of hyoscyamine 6-hydroxylase, a 2-oxoglutarate-dependent dioxygenase, from cultured roots of *Hyoscyamus niger*. J Biol Chem 266:9460–9464

Oksman-Caldentey K-M, Arroo R (2000) Regulation of tropane alkaloid metabolism in plants and plant cell cultures. In: Verpoorte R, Alfermann AW (eds) Metabolic engineering of plant secondary metabolism. Springer, New York, pp 253–281

Olmstead RG, Michaels HJ, Scott KM, Palmer JD (1992) Monophyly on the Asteridae and identification fron DNA sequences of rbcL. Ann Mo Bot Gard 79:249–265

Olmstead RG, Bremer B, Scott KM, Palmer JD (1993) A parsimony analysis of the Asteridaesensulato based on rbcL sequences. Ann Mo Bot Gard 80:700–722

Olmstead RG, Bohs L, Migid HA, Santiago-Valentini E, Garcia VF, Collier SM (2008) A molecular phylogeny of the Solanaceae. Taxon 54:1159–1181

Pagare S, Bhatia M, Tripathi N, Pagare S, Bansal YK (2015) Secondary metabolites of plants and their role: overview. Cur Trends Biotechnol Pharm 9(3):293–304

Palazón J, Navarro-Ocaña A, Hernandez-Vazquez L, Mirjalili MH (2008) Application of metabolic engineering to the production of scopolamine. Molecules 13:1722–1742. https://doi.org/10.3390/molecules13081722

Parr AJ (1992) Alternative metabolic fates of hygrine in transformed root cultures of *Nicandra physaloides*. Plant Cell Rep 11:270–273

Pigatto AGS, Blanco CC, Mentz LA, Soares GLG (2015) Tropane alkaloids and calystegines as chemotaxonomic markers in the Solanaceae. Anais Acad Bras Ciên 87(4):2139–2149

Pitta-Alvarez SI, Spollansky TC, Giulietti AM (2000) The influence of different biotic and abiotic elicitors on the production and profile of tropane alkaloids in hairy root cultures of *Brugmansia candida*. Enzym Microb Technol 26:252–258

Platonova TF, Kusovkov AD (1963) Med. Ind. UdSSR 17, 19 via Romeike, A (1978) Bot. Notiser, 131, 85

Pudersell K, Vardja T, Vardja R, Matto V, Arak E, Raal A (2012) Inorganic ions in the medium modify tropane alkaloids and riboflavin output in Hyoscyamus niger root cultures. Pharmacogn Mag 8(29):73–77. https://doi.org/10.4103/0973-1296.93330

Rivier L (1981) Analysis of alkaloids in leaves of cultivated *Erythroxylum* and characterization of alkaline substances used during coca chewing. J Ethnopharmacol 3:313–335

Robins RJ, Parr AJ, Payne J, Walton NJ, Rhodes MJC (1990) Factors regulating tropane-alkaloid production in a transformed root culture of a *Datura candida* x *D. aurea* hybrid. Planta 181:414–422

Robins RJ, Bachmann P, Peerless ACJ, Rabot S (1994) Esterification reactions in the biosynthesis tropane alkaloids in transformed root cultures. Plant Cell Tissue Organ Cult 38:241–247

Ryad A, Lakhdar K, Majda KS, Samia A, Mark A, Corrine AD, Eric G (2010) Optimization of the culture medium composition to improve the production of hyoscyamine in elicited *Datura stramonium* L. hairy roots using the response surface methodology (RSM). Int J Mol Sci 11:4726–4740

Samuelsson G (1999) Drugs of natural origin. A textbook of pharmacognosy, 4th edn. Swedish Pharmaceutical Press, Stockholm, pp 438–448

San Martin AS, Labbe C, Munoz O, Castillo M, Reina M, de la Fuente B, Gonzalez A (1987) Phytochemistry 26:819

Schulz OE (1907) Erythroxylaceae, vol 29. Engelmann, Leipzig

Seigler DS (1977) Plant systematics and alkaloids. In: Manske RHF (ed) The alkaloids, vol 16. Academic, New York, pp 1–82

Sharova EG, Aripova SF, Yunusov SY (1980) Alkaloids of Convolvulus subhirsutus. Khimiya Prirodnykh Soedinenii 5:672–676

Shoji T, Yamada Y, Hashimoto T (2000) Jasmonate induction of putrescine N-methyltransferase genes in the root of *Nicotiana sylvestris*. Plant Cell Physiol 41:831–839

Smith PM (1976) The chemotaxonomy of plants. Edward Arnold, London

Srivastava V, Mehrotra S, Mishra S (2018) Hairy roots: an effective tool of plant biotechnology. Springer, New York

Srivastava V, Mehrotra S, Mishra S (2020) Hairy root cultures based applications: methods and protocols. Springer, New York

Taktajan A (1997) Diversity and classification of flowering plants. Columbia University Press, New York

Tétényi P (1987) A chemotaxonomic classification of the Solanaceae. Ann Mo Bot Gard 74:600–608

Thorne RF (1992) Classification and geography of the flowering plants. Bot Rev 58:225–348

Waller GR, Nowachi EK (1978) Sites of alkaloid formation. In: Alkaloid biology and metabolism in plants. Plenum Press, New York, pp 121–141

Wang P, Yao T, Chen Z (1990) ZhiwuZhongyaoZazhi, 14: 739 (Ch.), via C.A., 113, 37699

Yao T, Chen Z, Yi D, Xu G (1981) YaoxueXuebao, 16, 582 (Ch.), via C.A., 96: 48972

Zerega NJC, Nur Supardi MN, Motley TJ (2010) Phylogeny and recircumscription of Artocarpeae (Moraceae) with a focus on *Artocarpus*. Syst Bot 35:766–782. https://doi.org/10.1600/036364410X539853

Zhang L, Yang B, Lu B, Kai G, Wang Z, Xia Y, Ding R, Zhang H, Sun X, Chen W, Tang K (2007) Tropane alkaloids production in transgenic *Hyoscyamus niger* hairy root cultures overexpressing putrescine N-methyltransferase is methyl jasmonate-dependent. Planta 225:887–896

Zhao T, Wang SLJ, Zhou Q, Yang C, Bai F, Lan X, Chen M, Liao Z (2020) Engineering tropane alkaloid production based on metabolic characterization of ornithine decarboxylase in *Atropa belladonna*. ACS Synth Biol 9(2):437–448

Zuanazzi J´AS, Tremea V, Limberger RP, Sobral M, Henriques A´ l T (2001) Alkaloids of Erythroxylum (Erythroxylaceae) species from Southern Brazil. Biochem Syst Ecol 29:819–825

Structure and Function of Enzymes Involved in the Biosynthesis of Tropane Alkaloids

2

Neill Kim, Benjamin Chavez, Charles Stewart Jr., and John C. D'Auria

Abstract

Tropane alkaloids are found in a scattered distribution among the angiosperm families including members within the Solanaceae, Erythroxylaceae, Convolvulaceae, and Brassicaceae. Recent studies regarding the origins of tropane production provide strong evidence for a polyphyletic origin, suggesting that novel enzymes from different gene families have been recruited during the course of flowering plant evolution. Tropane alkaloid biosynthesis is best documented on the molecular genetic and biochemical level from solanaceous species. Regardless of the system chosen, there are currently gaps in the knowledge of enzyme structure-function relationships and how they influence tropane alkaloid biosynthesis. Obtaining insights on structure-function relationships of tropane biosynthetic enzymes is critical to understanding regulation, turnover, and flux of metabolites through the pathway. In this review, we discuss the current state of knowledge regarding structure-function relationships of the known steps involved in tropane biosynthesis.

N. Kim
Department of Chemistry & Biochemistry, Texas Tech University, Lubbock, TX, USA
e-mail: neill.kim@ttu.edu

B. Chavez · J. C. D'Auria (✉)
Department of Molecular Genetics IPK Gatersleben, Leibniz Institute of Plant Genetics and Crop Plant Research, Gatersleben, Germany
e-mail: chavez@ipk-gatersleben.de; dauria@ipk-gatersleben.de

C. Stewart Jr.
Office of Biotechnology, Iowa State University, Ames, IA, USA
e-mail: cstewart@iastate.edu

V. Srivastava et al. (eds.), *Tropane Alkaloids*,
https://doi.org/10.1007/978-981-33-4535-5_2

Keywords

Erythroxylum coca · *Atropa belladonna* · Tropinone synthesis · Polyketide synthase · Tropane biosynthesis · Cocaine synthase

2.1 Introduction

Tropane alkaloids (TA) are plant specialized metabolites that have evolved as a response to nature's biotic and abiotic forces. Defined by their *N*-methyl-8-azabicyclo[3.2.1]-octane core structure, tropane alkaloid biosynthesis has been the subject of study for over a century due to its potent pharmacological activities (Fig. 2.1). There have been over 200 unique structures reported in the literature. These structures appear within seven different orders among angiosperms, which include ten different plant families. The families contributing the largest diversity of structures are the Solanaceae, Erythroxylaceae, Convolvulaceae, and Brassicaceae (Ziegler and Facchini 2008; Kim et al. 2016; Drager 2004; Lounasmaa and Tamminen 1993). The noncontiguous scattered distribution of tropane alkaloid-producing plant families gives rise to the question of whether or not tropane alkaloid biosynthesis is monophyletic or polyphyletic. The molecular data for the genes and enzymes responsible for tropane alkaloid biosynthesis was only available for members within the Solanaceae, until recently. Current data from the species *Erythroxylum coca* suggests that tropane alkaloid biosynthesis has arisen at least twice during the evolution of angiosperms (Jirschitzka et al. 2012).

Erythroxylum coca is one of the first domesticated plant species that was used for medicinal purposes, with evidence of its cultivation dating back at least 8000 years ago (Dillehay et al. 2010). The common occurrence of the carboxylic acid methyl ester located at the C2 position of cocaine is one of the most distinguishing characteristics of tropane alkaloids found in the Erythroxylaceae family. By binding the 3β benzoic ring present in cocaine to specific receptor sites, the reuptake of norepinephrine, serotonin, and dopamine is blocked and disrupts the normal physiology of the central nervous system. The 3β stereospecific conformation of tropane alkaloids is dominant in the Erythroxylaceae but only makes up a small component of tropane alkaloids found in members of the Solanaceae. The methylated nitrogen in the bicyclic core scaffold of cocaine and other tropane alkaloids serves as a structural analog of acetylcholine (Wink 1998a). Anticholinergics are a class of drugs used to block the action of the acetylcholine neurotransmitter in order to treat diseases such as Alzheimer's and Parkinson's and to alleviate motion sickness. Tropane alkaloids

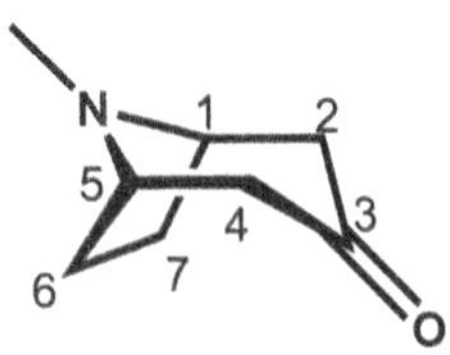

Fig. 2.1 Tropane alkaloid core bicyclic scaffold

have been detected to attach and inhibit the muscarinic acetylcholine receptors (Schmeller et al. 1995).

Solanaceous tropane alkaloids are also known for their anticholinergic and antispasmodic properties that affect the parasympathetic nervous system and have also been used for the treatment of anesthesia, for pain relief, and for drug addiction mediation (Shakeran et al. 2015; Xia et al. 2016; Qiang et al. 2016). The Solanaceae is the largest family known to produce tropane alkaloids, with 29 genera able to produce these metabolites. Atropine and scopolamine are well-known tropane alkaloids found in this family. In 1833, atropine was first isolated from *Atropa belladonna* (Geiger and Hesse 1833; Mein 1833). After much deliberation and structural studies, the correct structure of atropine was obtained in 1889 (Wink 1998b). Scopolamine is commonly found in the plant *Hyoscyamus niger* and capable of crossing the blood-brain barrier to affect the central nervous system (Wink 1998b). However, due to the low metabolite levels found in plants, there is an ongoing effort to thoroughly understand the biosynthesis of scopolamine and similar metabolites.

With the current knowledge gained regarding these valued pharmaceutical compounds, there is an increasing interest in the elucidation of their biosynthetic pathway(s) in order to upregulate TA production. Moreover, the World Health Organization (WHO) continues to include these pharmaceutically important tropane alkaloid metabolites on their list of essential drugs (WHO 2019). Tropane alkaloid levels in plants are generally present in low quantities during normal plant biosynthesis. Chemically synthesizing these alkaloids in the laboratory has also demonstrated to be laborious and costly due to their important stereochemical nature. There have been attempts at a total synthesis approach to synthesize scopolamine, but low yields of only 16% did not make this method economically or environmentally practical (Nocquet and Opatz 2016). The leading complication for the commercial production of scopolamine in hairy root cultures is attaining industrial level yields (Ullrich et al. 2016). For this reason, researchers are focusing their efforts on elucidating and understanding tropane alkaloid biosynthesis in all known plant families for future metabolic engineering efforts.

2.2 TA Core Biosynthesis

The elucidation of structures and potential biosynthetic steps in the tropane alkaloid pathway have predominantly used methods such as radioisotope-labeled feeding studies that are followed by chemical degradation analysis. Primarily, tropane alkaloid biosynthesis begins with the recruitment of amino acids from primary metabolism into a nitrogen-containing heterocyclic ring intermediate (Scheme 2.1). This heterocycle will then continue to form the second ring in the tropane alkaloid bicyclic scaffold, which is finally followed by modifications through the addition of diverse functional groups yielding the final compound. The starting substrates for tropane alkaloid biosynthesis were predicted to be ornithine and arginine as early as 1954 (Leete et al. 1954). Feeding studies performed on the

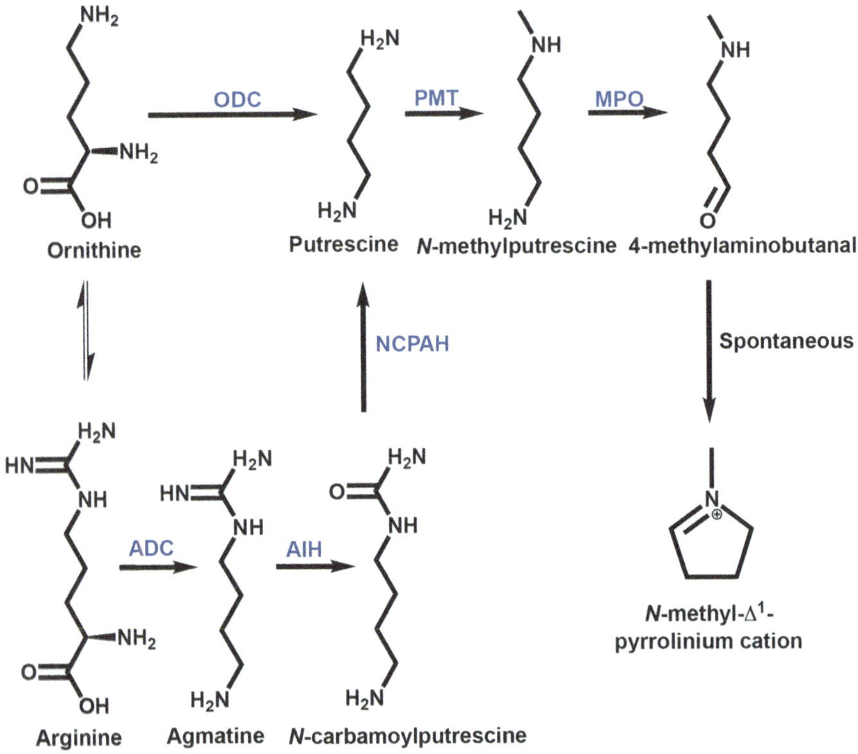

Scheme 2.1 The initial steps of tropane alkaloid biosynthesis up to the formation of *N*-methyl-Δ^1-pyrrolinium. The enzymes depicted in blue are as follows: arginine decarboxylase (ADC), agmatine iminohydrolase (AIH), *N*-carbamoylputrescine amidohydrolase (NCPAH), ornithine decarboxylase (ODC), putrescine *N*-methyltransferase (PMT), and methylputrescine oxidase (MPO)

roots of *Atropa belladonna* using ^{14}C-proline proposes the possible incorporation of the amino acid proline into the tropane ring (Soeren 1962). Furthermore, studies using *Datura stramonium* and *Datura metel* have also reported the incorporation of proline into the tropane alkaloid compounds tropine and scopolamine (Liebisch and Schütte 1967). The commonly shared intermediate, pyrroline-5-carboxylate, links arginine, ornithine, and proline together. These three amino acids are readily interconvertible making radiolabeled amino acid feeding studies challenging to interpret without further enzymological data (Delauney and Verma 1993).

It has been hypothesized that a nonsymmetrical intermediate is involved in the production of the pyrrolidine ring if ornithine is first methylated at the δ-*N* position. An alternative route proposes that ornithine undergoes a decarboxylation to form the polyamine putrescine as the first biosynthetic step (Leete 1990, 1962). Radiolabeled ornithine-2-^{14}C was fed to several different *Datura* plant species and showed the incorporation of a nonsymmetrical intermediate. In contrast, the *Nicotiana*, *Erythroxylum*, and *Hyoscyamus* plant species have reported a symmetrical intermediate whose incorporation showed activity at positions C1 and C5 of the tropane ring (Leete 1982). Ornithine can be converted into the symmetrical intermediate putrescine through a one-step enzymatic reaction facilitated by ornithine decarboxylase (ODC, EC 4.1.1.17), an enzyme that has been isolated in several tropane alkaloid-producing plant species and is a pyridoxal phosphate-dependent decarboxylase (Docimo et al. 2012; Imanishi et al. 1998; Michael et al. 1996). ODC has been predicted to be a cytosolic enzyme but seems to accumulate in the nucleus (Sandmeier et al. 1994; Schipper et al. 2004). Malmberg et al. reveal that putrescine produced by ODC is important for metabolic processes such as cellular differentiation, development, and division (Malmberg et al. 1998). To date, the only enzymatically characterized ornithine decarboxylase enzymes from plants are from *N. glutinosa* (NgODC), *E. coca* (EcODC), and *H. niger* (HnODC) (Zhao et al. 2019). Currently there are no crystal structures of a plant ODC. However, in 2001, Lee and Cho created the first model of a plant ornithine decarboxylase from *N. glutinosa* (Lee and Cho 2001). The three-dimensional model of NgODC is based on the tertiary crystal structure of both a mouse ODC and an ODC from *Trypanosoma brucei*. This model predicts NgODC to be a symmetrical homodimer. Most ODCs have a PFYAVKCN and a GPTCD binding motif for the pyridoxal 5′-phosphate (PLP) cofactor (Zhao et al. 2019; Coleman et al. 1993). PLP-dependent decarboxylases require a conserved lysine residue in order to form a Schiff base and to stabilize the binding of the PLP to the active site. The PLP binding site of NgODC is located on the C-terminal end of the *α/β* barrel. The active site contains Lys95, Cys96, Cys338, and Cys377 as key amino acid residues, and mutagenesis experiments on the Lys95 residue to alanine in NgODC showed a significant decrease in the catalytic efficiency. The Cys377 residue has also been implicated as the key residue for the covalent binding of the DL-α-difluoromethylornithine (DFMO) suicide inhibitor (Lee and Cho 2001; Coleman et al. 1993).

There is also a more indirect route to putrescine that begins with the amino acid arginine. This three-step pathway to putrescine begins with the decarboxylation of arginine, via arginine decarboxylase (ADC; EC 4.1.1.19) using PLP as the cofactor, to form agmatine (Docimo et al. 2012). ADC is typically localized in the chloroplasts of plants but has also been found in other regions such as the mitochondria and the cytosol (Bortolotti et al. 2004; Slocum et al. 1984; Borrell et al. 1995). Putrescine derived from ADC is thought to be associated with nondividing tissues or tissues responding to environmental stresses (Docimo et al. 2012). Little structural data of ADC exists, and there are no crystal structures of a plant arginine decarboxylase. ADC is suggested to be a trimer in soybeans and oats (Nam et al. 1997; Malmberg et al. 1992). *E. coca* ADC (EcADC) contains a PLP binding site and a decarboxylase binding motif (Docimo et al. 2012). Much like ODC, ADC requires a conserved lysine residue for PLP binding. Docimo et al. reported on the characterization of ADC and ODC in *E. coca* but could not determine which enzyme is primarily involved in tropane alkaloid biosynthesis.

The decarboxylated product of ADC, agmatine, is then converted into *N*-carbamoylputrescine via agmatine iminohydrolase (AIH; EC 3.5.3.12). AIH belongs to the *Porphyromonas*-type peptidylarginine deiminase family which is part of the penteins superfamily (Hartzoulakis et al. 2007). A notable characteristic of the penteins superfamily includes a propeller-shaped protein fold which forms a narrow channel in the core where substrate binding occurs. Currently, there are no characterized AIH enzymes in the tropane-producing members of the Solanaceae or Erythroxylaceae family. Putrescine production within solanaceous plants predominantly relies on the decarboxylation of ornithine as the primary source of putrescine for tropane alkaloid metabolism. However, feeding studies using radiolabeled agmatine were shown to be incorporated into hyoscyamine, providing evidence that hyoscyamine may be derived from arginine instead of ornithine in *H. niger* root cultures (Walton et al. 1990). The active site of AIH was speculated to use a cysteine residue to form a thioester with the agmatine substrate, but mutagenesis experiments of AIH (At5g0817) cloned from *A. thaliana* revealed two conserved cysteine residues that are important but not essential for enzymatic activity (Janowitz et al. 2003). Although there are no crystal structures of AIH in members of the Solanaceae or Erythroxylaceae, AIH is reported to be homodimer in *Z. mays*, *A. thaliana*, and *O. sativa* (Janowitz et al. 2003; Yanagisawa and Suzuki 1981; Mohan Chaudhuri and Ghosh 1985). Recently, a crystal structure of AIH has been reported for *Medicago truncatula* (MtAIH) (Sekula and Dauter 2019a). The crystal structure of MtAIH portrays a symmetrical homodimer with two subunits arranged in a propeller fold geometry with five αββαβ repeated units.

Finally, *N*-carbamoylputrescine is then converted into putrescine via *N*-carbamoylputrescine amidohydrolase (NCPAH; EC 3.5.1.53) to complete the alternative three-step pathway to putrescine. An NCPAH enzyme was found in *A. thaliana* with kinetic data that reveals a sigmoidal curve, indicating positive cooperativity (Piotrowski et al. 2003). Investigations into this cooperativity revealed an accelerating mechanism, activated by saturating amounts of *N*-carbamoylputrescine, and determined that AtNCPAH is not a rate-limiting step in putrescine biosynthesis. Much like AIH, NCPAH has not been well characterized in tropane alkaloid biosynthesis. This lack of characterization limits current knowledge on the structure-function relationship within tropane alkaloid-producing plants. However, a recent report on the crystal structure of NCPAH in *M. truncatula* (MtNCPAH) shows helical octamers with a funnel-shaped active site containing glutamate as a proton acceptor, lysine as a proton donor, and cysteine as a nucleophile (Sekula et al. 2016). The catalytic residues of MtNCPAH are located deep within the narrow section of the active site. MtNCPAH is notably substrate specific for *N*-carbamoylputrescine due to a negatively charged glutamate residue located at the entrance of the active site. This glutamate residue determines the size and length of substrates allowed to bind within the active site. It could not be ascertained if allosteric regulation plays a role in MtNCPAH as previously shown in *A. thaliana*. Molecular docking experiments of MtNCPAH suggest that a secondary binding site allows for an additional *N*-carbamoylputrescine to be bound before entering the active site. This would allow for an increase in the enzyme turnover rate due to a decrease in time needed for the substrate to diffuse into the active site; however, this hypothesis has yet to be experimentally confirmed.

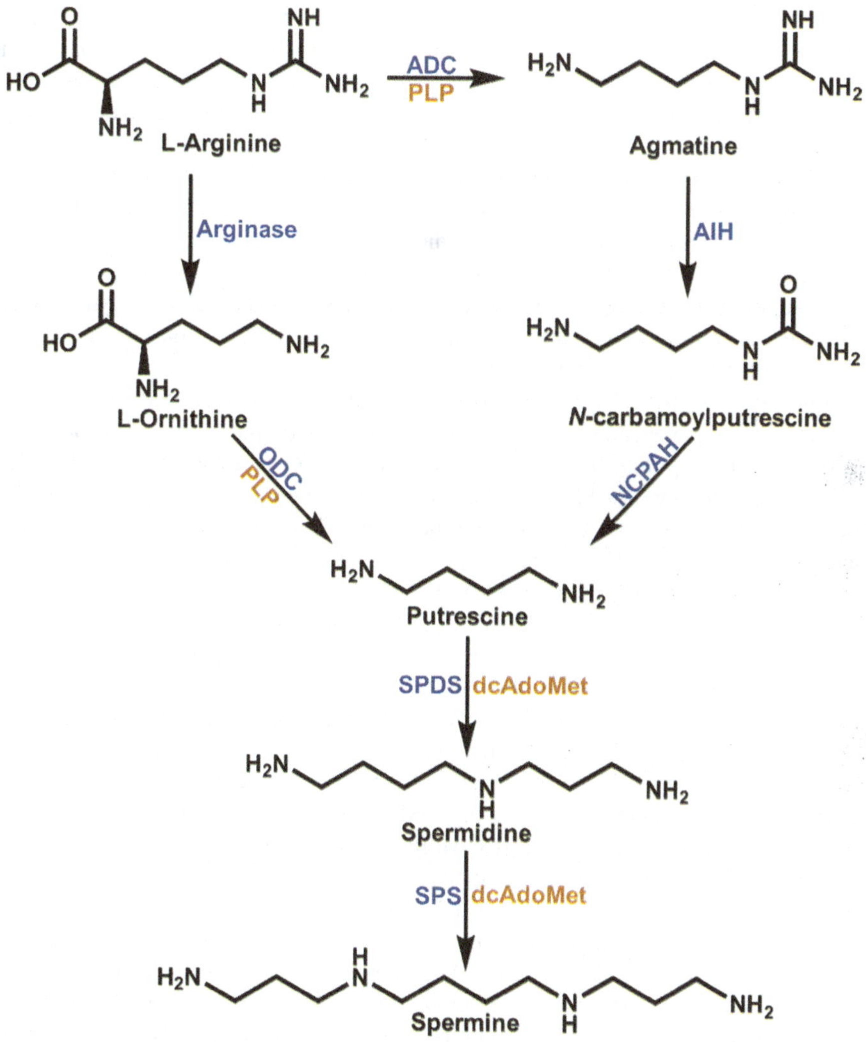

Scheme 2.2 The biosynthetic pathway of polyamines in plants. Enzymes depicted in blue are listed as follows: arginine decarboxylase (ADC), agmatine iminohydrolase (AIH), *N*-carbamoylputrescine amidohydrolase (NCPAH), ornithine decarboxylase (ODC), spermidine synthase (SPDS), and spermine synthase (SPS). The cofactors depicted in orange are pyridoxal phosphate (PLP) and decarboxylated *S*-adenosylmethionine (dcAdoMet)

In primary metabolism, the polyamine serves as a precursor for other polyamines such as spermidine and spermine (Scheme 2.2). These compounds are involved in both primary and secondary metabolism in plants. The most common polyamines are putrescine, spermidine, spermine, and cadaverine which are derived from amino acids such as arginine, ornithine, and lysine (Miller-Fleming et al. 2015). Polyamines are a vital class of nitrogen-rich molecules that are found in all kingdoms of life and participate in a wide variety of biochemical functions. The small size of these compounds and their cationic nature allow for multiple interactions with cellular components such as DNA, RNA, phospholipids, proteins, and chromatin (Kusano et al. 2008). Due to these broad cellular interactions, polyamines are implicated in a wide array of essential cellular processes such as gene expression, translation, signaling, and membrane stability (Tabor and Tabor 1984). Polyamine-deficient *Escherichia coli* shows increased toxicity to oxygen and increased hyper-sensitivity to hydrogen peroxide implying that these metabolites play a role in buffering reactive oxygen species (Chattopadhyay et al. 2003). In plants, polyamines are accumulated due to biotic and abiotic stresses. Exogenously added polyamines to salt stressed *Oryza sativa* plants gave indications of polyamine's protective role for abiotic stress responses (Chattopadhayay et al. 2002).

Putrescine *N*-methyltransferase (PMT) catalyzes the committed step toward tropane alkaloid biosynthesis, the generation of *N*-methylputrescine (Biastoff et al. 2009a). *N*-methylputrescine is formed from the transfer of a methyl group from *S*-adenosylmethionine (SAM, AdoMet) onto an amino group of putrescine. As described above, putrescine is a central metabolite for the production of spermidine and other polyamines. The species distribution of PMT appears to be taxonomically restricted to nicotine and tropane alkaloid-producing plant species (e.g., Solanaceae, Convolvulaceae) (Biastoff et al. 2009a). PMT is biochemically similar to spermidine synthase (SPDS), a ubiquitous enzyme in plants, animals, and bacteria involved in polyamine biosynthesis. While both SPDS and PMT use putrescine as a substrate, SPDS does not accept AdoMet as a methyl donor. Rather, SPDSs use decarboxylated SAM (dcAdoMet) to transfer an aminopropyl group onto putrescine. The biochemical similarities and mutational studies discussed above are consistent with phylogenetic analyses indicating PMTs evolved from SPDSs via a gene duplication event (Junker et al. 2013; Minguet et al. 2008).

The binding pockets of PMTs appear to be similar to SPDSs. Unfortunately, X-ray diffraction studies of PMT crystals have been unsuccessful to date (Biastoff et al. 2009a). However, crystal structures of SPDSs, including the recently published structures of two SPDS isoforms from *Arabidopsis*, have provided structural insights into the function of PMTs (Sekula and Dauter 2019b). Homology models indicate that the overall structure of PMT is similar to SPDS (Junker et al. 2013; Teuber et al. 2007). Similar to most SPDSs, PMTs appear to be dimeric with each monomeric subunit containing an independent active site (Sekula and Dauter 2019b; Biastoff et al. 2009b). The active site of PMT/SPDS is formed via a cleft between the N-terminal domain and the C-terminal domain of each monomer. Some PMTs (i.e., PMTs from *Datura* spp. and *Nicotiana* spp.) contain an 11-amino-acid tandem repeat on their N-terminus experimentally observed to affect protein solubility but not catalytic activity (Biastoff et al. 2009a; Hashimoto et al. 1998). However, chimeras generated from swapping N- and C-terminal portions of PMT and SPDS from *D. stramonium* revealed reaction specificity was largely determined by the residues in the N-terminal portion of the molecules (Biastoff et al. 2009b). The C-terminal portion of SPDS contains a Rossmann-like fold which may be involved in substrate binding (Junker et al. 2013; Sekula and Dauter 2019b).

The divergence of PMT from SPDS likely required changes in the conformation of the methionine moiety of SAM relative to the aminopropyl group of dcSAM. The SN2 reaction catalyzed by PMT and SPDS requires a short distance between the donor group, either methyl or aminopropyl, respectively, and a correctly oriented amino group of the acceptor molecule, putrescine. In 2009, Biastoff et al. initially hypothesized that PMT and SPDS had different binding locations for putrescine (Biastoff et al. 2009b). Junker et al. discussed other possibilities including inversion of the sulfur chirality of SAM as well as conformational changes of the methionine moiety of SAM (Junker et al. 2013). Ab initio calculations of the inversion barrier led this group to dismiss the possibility of a switch of the sulfur chirality of SAM (*S*-SAM to *R*-SAM). Comparative structural analyses of a PMT homology model with apo and ligand-complexed SPDS crystal structures indicated that residues binding the adenosyl moiety of SAM/dcSAM are highly conserved. However, Junker and coworker's comparative analysis identified several mutations that putatively alter the conformation of the (decarboxy)methionine moieties of SAM/dcSAM. Results from mutagenesis experiments by Junker et al. support that only a few mutations, likely affecting the conformation of the methionine moiety of SAM, were necessary for PMTs to diverge from SPDSs.

N-methylputrescine must then undergo an oxidative deamination reaction in order to form the cyclic pyrrolidine ring. 4-Methylaminobutanal is the resulting product that undergoes spontaneous cyclization to yield the *N*-methyl-Δ^1-pyrrolinium cation (Leete 1990). Amine oxidases catalyze the oxidative deamination of various amines into aminoaldehydes and H_2O_2 and are classified into the FAD-dependent polyamine oxidases (PAOs) and the copper-containing amine oxidases (CuAOs) (Naconsie et al. 2014). These enzymes have many differences to each other such as substrate specificity, catalytic mechanism, subcellular localization, and functional diversity (Tavladoraki et al. 2016). The PAOs in plants have a non-covalently bound FAD molecule as a cofactor and are known to oxidize the secondary amino group of polyamines, with the reaction products dependent on the catalytic mechanism and substrate specificity. All intracellular PAOs oxidize the carbon located at the *exo*-side of the N4 atom of spermine or spermidine to produce 3-aminopropanal, H_2O_2, and spermidine or putrescine, respectively (Tavladoraki et al. 2006; Kamada-Nobusada et al. 2008; Moschou et al. 2008; Fincato et al. 2011; Ahou et al. 2014; Kim et al. 2014; Liu et al. 2014a; Mo et al. 2015). Conversely, the apoplastic PAOs are known to oxidize the carbon located at the *endo*-side of the N4 atom of spermine and spermidine to produce aminoaldehyde, H_2O_2, and 1,3-diaminopropane (Tavladoraki et al. 1998; Cervelli et al. 2001; Liu et al. 2014b). The CuAOs predominantly oxidize the aliphatic diamines putrescine and cadaverine to produce an aminoaldehyde, ammonia, and H_2O_2 and are considered to be involved in polyamine terminal catabolism (Tavladoraki et al. 2016). They are also capable of oxidizing the primary amino groups of spermine and spermidine but less efficiently. Plant CuAOs are homodimeric enzymes where each subunit contains a copper ion to oxidize the conserved tyrosine residue located in the catalytic site into the cofactor 2,4,5-trihydroxyphenylalanine quinone, also known as topoquinone (Naconsie et al.

2014). CuAOs also appear at high levels in dicots, especially in pea, chickpea, lentil, and soybean seedlings, while PAOs are highly expressed, particularly in monocots (Cona et al. 2006; Šebela et al. 2001).

In solanaceous plants, methylputrescine oxidase (MPO) catalyzes the next step in tropane alkaloid biosynthesis through oxidative deamination by taking *N*-methylputrescine and converting it into 4-methylaminobutanal. The first MPO to be characterized was from the species *Nicotiana tabacum* and belongs to the CuAO class of diamine oxidases (Heim et al. 2007; Katoh et al. 2007; Matsuzaki et al. 1994). In *D. stramonium*, labeled 4-methylaminobutanal were detected in plants that were fed with [2-^{14}C]-ornithine (Leete 1990). Several other tropane alkaloid-producing members in the Solanaceae have reported enzyme activities containing this type of oxidation reaction (Feth et al. 1985; Hashimoto et al. 1990; Mizusaki et al. 1973). In nicotine biosynthesis, evidence suggests MPO affiliates with other important enzymes leading to the hypothesis that a metabolic channel exists in which a multienzyme complex is active (Leete 1990). Currently there are no MPO-like genes to be discovered in the Erythroxylaceae family. However, the same enantioselectivity found in solanaceous plants that produce either tropane alkaloids or nicotine was shown to have analogous activity when intact plants were fed 4-monodeuterated *N*-methylputrescine (Hoye et al. 2000). Through ^{13}C fractionation techniques, researchers were able to establish that both nicotine and hyoscyamine share the same biosynthetic pathway up to the *N*-methyl-Δ^1-pyrrolinium cation (Romek et al. 2016). These data strongly suggest that there is an MPO homolog in the Erythroxylaceae family (Wigle et al. 1982). However, there has been evidence shown that the *N*-methyl-Δ^1-pyrrolinium cation could be indirectly derived from spermidine in nicotine biosynthesis in *Nicotiana glutinosa* (Leete 1985). 4-Methyaminobutanal then spontaneously cyclizes to form the *N*-methyl-Δ^1-pyrrolinium cation which serves as the first ring in the bicyclic tropane core scaffold (Leete 1990). In *E. coca*, feeding studies using [2-^{13}C, ^{15}N]-*N*-methylpyrrolinium chloride into coca leaves demonstrated the incorporation of the *N*-methyl-Δ^1-pyrrolinium cation into methylecgonine, a precursor to the tropane alkaloid cocaine (Leete et al. 1991, 1988a). This is also the first occurrence of a spontaneous reaction in the tropane alkaloid pathway.

Since the chemical synthesis of tropinone and the earliest investigations of tropane biosynthesis nearly a century ago, the second ring formation in tropane alkaloids has been a mystery (Scheme 2.3). Isotope labeling studies indicated that the second ring appeared to form via the condensation of acetate units onto the *N*-methyl-Δ^1-pyrrolinium cation (Kaczkowski et al. 1961; Leete 1983; Liebisch et al. 1972). One molecule that received a lot of attention as a potential intermediate of the second ring formation was hygrine, (*R*)-1-(1-methylpyrrolidin-2-yl)-propan-2-one. Early isotope labeling studies indicated that hygrine was likely formed from the spontaneous decarboxylation of a β-keto acid produced from the enzymatic condensation of acetoacetic acid or acetoacetyl-CoA onto the *N*-methyl-Δ^1-pyrrolinium cation. Hygrine as an intermediate lost favor when it was found that the condensation between *N*-methyl-Δ^1-pyrrolinium cation and acetoacetyl-CoA can occur nonenzymatically (Endo et al. 1988). Additionally, other studies have demonstrated

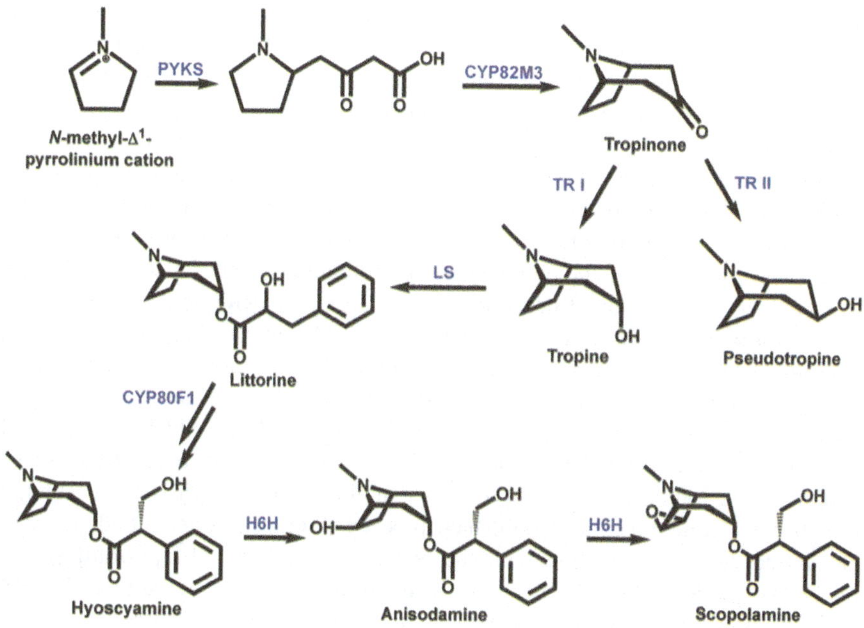

Scheme 2.3 Biosynthesis of scopolamine in the Solanaceae family starting from the *N*-methyl-Δ^1-pyrrolinium cation. Enzymes depicted in blue are as follows: pyrrolidine ketide synthase (PYKS), cytochrome p450 (CYP82M3), tropinone reductase I (TR I), tropinone reductase II (TR II), littorine synthase (LS), cytochrome p450 (CYP80F1), and hyoscyamine-6β-hydroxylase (H6H). Double arrows depict a multistep reaction requiring another enzyme

that hygrine is an experimental artifact of another compound (Kaczkowski et al. 1961; Liebisch et al. 1972; Abraham and Leete 1995). However, other labeling studies using *Datura* spp. (Solanaceae) showed the incorporation of racemic ethyl [2,3-$^{13}C_2$]-4-(*N*-methyl-2-pyrrolidinyl)-3-oxobutanoate, a polyketide-derived compound, into the second ring structure (Abraham and Leete 1995; Robins et al. 1997). Additionally, feeding studies of methyl (RS)-[1,2-$^{13}C_2$,1-^{14}C]-4-(*N*-methyl-2-pyrrolidinyl)-3-oxobutanoate to the leaves of *E. coca* further demonstrated the involvement of a polyketide molecule (Leete et al. 1991).

Based on the feeding studies described above, the second ring formation in tropane alkaloid biosynthesis has been hypothesized to arise via the activity of a type III polyketide synthase (PKS) catalyzing the condensation of acetate units onto the *N*-methyl-Δ^1-pyrrolinium cation (Kim et al. 2016; Abraham and Leete 1995; Robins et al. 1997; Humphrey and O'Hagan 2001). Type III PKSs are promiscuous enzymes that have an extensive tolerance for diverse substrates while being able to catalyze multiple reactions (Austin and Noel 2003; Stewart Jr et al. 2013). The type III polyketide synthase family of enzymes are homodimeric proteins containing a Cys-His-Asn catalytic triad typically known to catalyze the binding of a CoA-tethered starter substrate, less frequently ACP-tethered, onto the catalytic cysteine, as reviewed in Austin and Noel (Austin and Noel 2003). Once covalently attached to the active site cysteine, the starter substrate undergoes multiple rounds of decarboxylative Claisen condensation, typically using malonyl-CoA as the second/extending substrate, to build a polyketide chain. Hydrolysis and/or cyclization of the elongated polyketide intermediate ends the chain-elongation process.

Recently, a major breakthrough in understanding the formation of the second ring occurred when two type III PKSs were independently isolated and experimentally confirmed to be involved in tropane alkaloid biosynthesis. First, Bedewitz et al. (2018) isolated a type III PKS from the transcriptome of *Atropa belladonna* roots (AbPYKS) that accepted the *N*-methyl-Δ^1-pyrrolinium cation as a starter substrate and catalyzed two rounds of decarboxylative Claisen condensations with malonyl-CoA to form the 4-(*N*-methyl-2-pyrrolidinyl)-3-oxobutanoic acid intermediate using in vitro assays (Bedewitz et al. 2018). Additional virus-induced gene silencing (VIGS) experiments confirmed that AbPYKS played a critical role in tropane alkaloid biosynthesis. Bedewitz and coworkers postulated that AbPYKS catalyzed the formation of a CoA-tethered intermediate via the decarboxylate condensation of malonyl-CoA onto the *N*-methyl-Δ^1-pyrrolinium cation. This CoA-tethered intermediate, a monoketide, would then proceed through another round of chain elongation before terminating with the release of 4-(*N*-methyl-2-pyrrolidinyl)-3-oxobutanoic acid, a diketide. Second, Huang et al. (2019) isolated a type III PKS from the transcriptomes of *Anisodus acutangulus* hairy root cultures (AaPYKS) whose initial in vitro activity also suggested that it directly produced 4-(*N*-methyl-2-pyrrolidinyl)-3-oxobutanoic acid (Huang et al. 2019) (Scheme 2.4). During their efforts to solve the crystal structure of AaPYKS, Huang et al. serendipitously trapped an acyl-enzyme intermediate (PDB ID: 6J1M) and a CoA-tethered intermediate (PDB ID: 6J1N) of AaPYKS. The acyl-enzyme intermediate consists of 4-carboxy-3-oxobutanoyl (COB) covalently bonded via a thioester linkage to the catalytic cysteine of AaPYKS. Critically, the carboxylic acid functional group of COB was prevented from spontaneously decarboxylating via a salt bridge and hydrogen bond with nearby arginine and serine side chains, respectively. Interestingly, when crystals of the acyl-enzyme intermediate (AaPYKS-COB) were soaked with malonyl-CoA, Huang and coworkers generated a non-covalent intermediate complex of AaPYKS with 4-carboxy-3-oxobutanoyl-CoA (Fig. 2.2). Further in vitro assays revealed that in the absence of *N*-methyl-Δ^1-pyrrolinium cations, AaPYKS generated 3-oxoglutaric acid (β-ketoglutarate) using malonyl-CoA as both a starter and extending substrate. Additionally, 3-oxoglutaric acid was observed by Huang et al. to spontaneously react with *N*-methyl-Δ^1-pyrrolinium cations to yield 4-(*N*-methyl-2-pyrrolidinyl)-3-oxobutanoic acid. Lastly, the stereochemistry of the enzymatically produced 4-(*N*-methyl-2-pyrrolidinyl)-3-oxobutanoic acid was determined to be a racemic mixture, consistent with the product of the spontaneous reaction of 3-oxoglutaric acid with the *N*-methyl-Δ^1-pyrrolinium cation (Huang et al. 2019). Collectively, the results of their structural and biochemical analysis allowed Huang et al. to deduce that AaPYKS, and likely AbPYKS, catalyze the formation of 3-oxoglutaric acid from two malonyl-CoA molecules. 3-Oxoglutaric acid subsequently reacts spontaneously with the *N*-methyl-Δ(Ziegler and Facchini 2008)-pyrrolinium cation to yield 4-(*N*-methyl-2-pyrrolidinyl)-3-oxobutanoic acid which

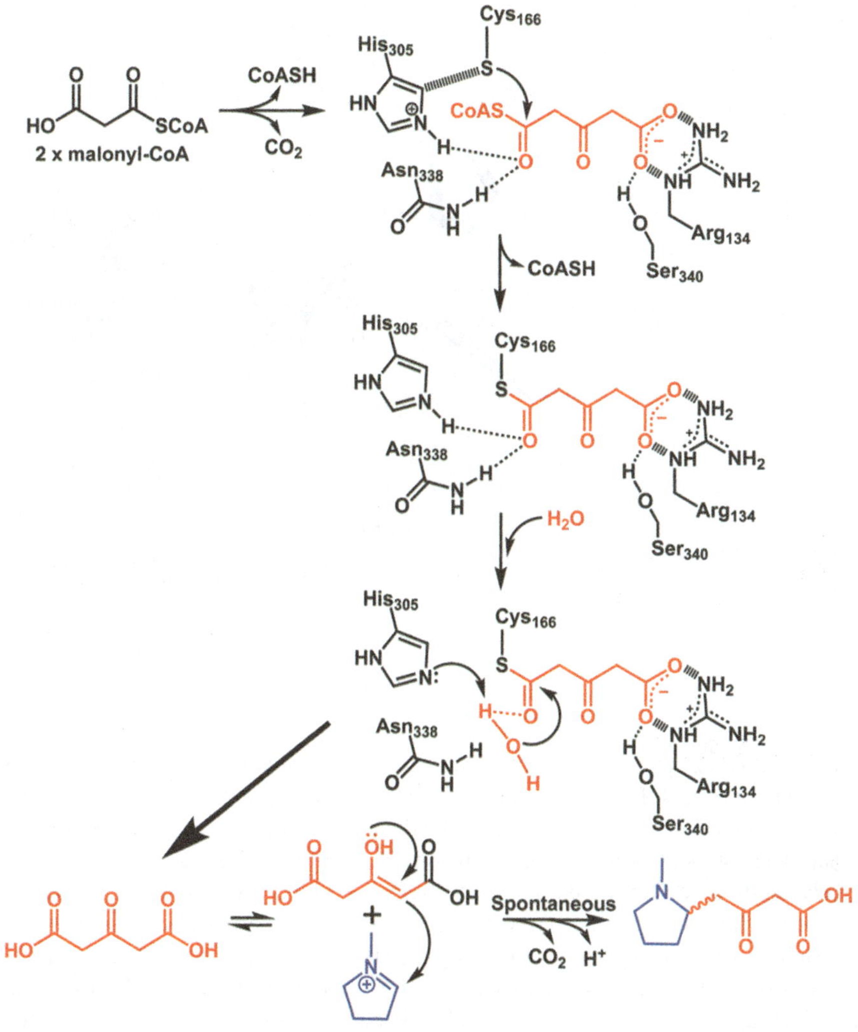

Scheme 2.4 Hypothetic mechanism of AaPYKS for 4-(*N*-methyl-2-pyrrolidinyl)-3-oxobutanoic acid formation

is ultimately cyclized by a cytochrome p450 (see below) to generate the second ring of the tropane scaffold.

Both PYKS mechanisms postulated by Bedewitz et al. (2018) and Huang et al. (2019) rely on a novel type III PKS activity. The amino acid sequences of AbPYKS and AaPYKS are 85% identical (92% similar), have conserved the catalytic triad found in all type III PKSs, and have conserved key amino acids observed by Huang et al. that interact with intermediates. In their discussion of AbPYKS activity, Bedewitz et al. postulated that *N*-methyl-Δ^1-pyrrolinium cations are directly used as a substrate by PYKSs. A charged and non-CoA/ACP tethered substrate such as

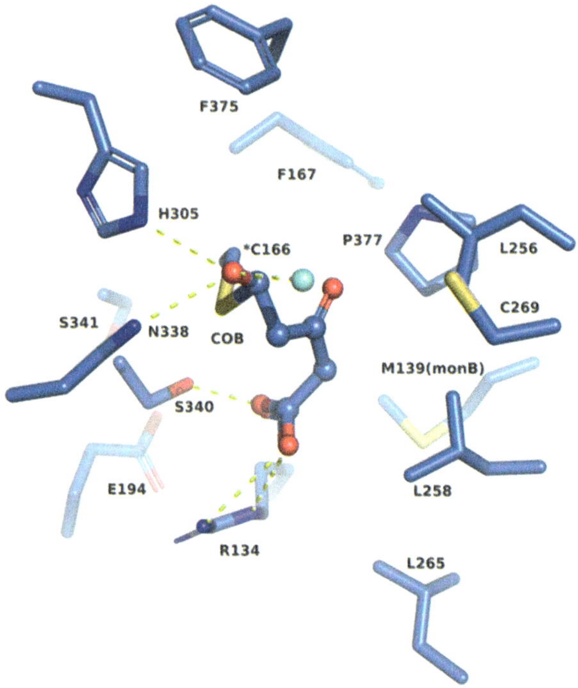

Fig. 2.2 Crystal structure of AaPYKS-COB looking down the CoA binding tunnel (2.0 Å). COB moiety depicted in ball and stick model

the N-methyl-Δ^1-pyrrolinium cation has never been reported as a natural starter substrate for type III PKSs, to the best of our knowledge. Conversely, the mechanism presented by Huang and coworkers involved PYKSs using an intact malonyl-CoA as a starter substrate. Given the pKa of malonyl-CoA (~5.0), this starter malonyl-CoA likely binds to PYKS as its conjugate base, a carboxylate anion. As mentioned above, a charged substrate has not been previously reported as a natural starter substrate for PKSs; neither has the biosynthesis of 3-oxoglutaric acid via a type III PKS. Several key questions about the PYKSs remain to be clarified. Given that *AbPYKS* and *AaPYKS* come from closely related species within the Solanaceae, will PYKSs from other tropane alkaloid-producing plants show similar biochemical function? Additionally, 3-oxoglutaric acid and 4-(*N*-methyl-2-pyrrolidinyl)-3-oxobutanoic acid are β-keto acids. A mechanism by which PYKS prevents undesirable spontaneous decarboxylation is unclear. *O*-methylation of 4-(*N*-methyl-2-pyrrolidinyl)-3-oxobutanoic acid would prevent decarboxylation (see below), but an *O*-methyltransferase for such a reaction has not been definitively characterized, and the sequence of events involving the activities of PYKS, *O*-methyltransferase, and cytochrome p450 responsible for the second ring cyclization is unclear. Lastly, the PYKS reaction scheme presented by Huang et al. juxtaposes a second spontaneous reaction in the tropane alkaloid biosynthesis. First, 4-methylaminobutanal spontaneously cyclized to form the N-methyl-Δ^1-pyrrolinium cation (see MPO section above), and then the N-methyl-Δ^1-pyrrolinium cation is a substrate for a

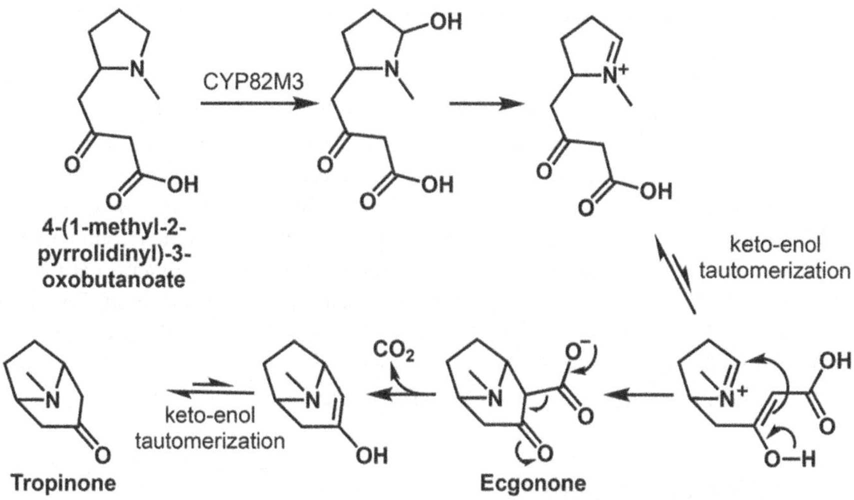

Scheme 2.5 Proposed mechanism of CYP82M3-catalyzed reaction

spontaneous condensation reaction with 3-oxoglutaric acid. Such a sequence of event warrants a close examination of the thermodynamic properties of tropane alkaloid biosynthesis.

The enzyme responsible for the cyclization of the 4-(*N*-methyl-2-pyrrolidinyl)-3-oxobutanoic acid intermediate to form the second ring structure of the tropane alkaloid core scaffold was determined to be a cytochrome p450 (AbCYP82M3) found in solanaceous plants, yielding tropinone (Bedewitz et al. 2018) (Scheme 2.5). The cytochrome p450 subfamily CYP82 are known to catalyze diverse reactions in plant specialized metabolism such as CYP82Y1 which catalyzes the first committed step in noscapine biosynthesis in opium poppy that takes *N*-methylcanadine to form 1-hydroxy-*N*-methylcanadine (Dang and Facchini 2014). AbCYP82M3 is closely related to several CPY82s found in tobacco that encode nicotine *N*-demethylases. These enzymes catalyze the formation of nornicotine from nicotine through a demethylation reaction on the pyrrolidine ring of nicotine (Siminszky et al. 2005; Lewis et al. 2010). The pyrrolidine ring is structurally similar to the ring found in the 4-(*N*-methyl-2-pyrrolidinyl)-3-oxobutanoic acid intermediate of AbCYP82M3, suggesting that these enzymes could share similar substrate binding domains and could possibly have evolved from a common ancestor. The structure-function analysis of the catalytic mechanism for this enzyme was not investigated in this study, but they do report a hypothetical mechanism that involves the restoration of an electrophilic iminium cation. It is a similar mechanism proposed for the formation of quinine from strictosidine, where iminium and aldehyde are intermediates for the quinidine and quinolone moieties (Kacprzak 2013). The same group hypothesized that AbCYP82M3 catalyzes the hydroxylation of the pyrrolidine ring at the C5 position to form a hydroxyl that would then undergo a dehydration reaction to produce a second iminium intermediate. The ketone in the 3-oxobutanoic acid moiety would undergo keto-enol tautomerization to yield a nucleophilic enol that allows for an intramolecular condensation reaction between the C3′ and C5 positions producing 2-carboxytropinone, also known as ecgonone. Finally, decarboxylation of the β-ketoacid would lead to tropinone. Although this is a conceivable model, other plausible models should not be ruled out. As stated earlier, methyl ester formation at the C2 position of 4-(*N*-methyl-2-pyrrolidinyl)-3-oxobutanoic acid intermediate can prevent spontaneous decarboxylation. This would also support the presence of a carbomethoxy group found in cocaine and other tropane alkaloids found in the Erythroxylaceae such as methylecgonone which contains a keto function at the C3 position and a carbomethoxy group at the C2 position (Jirschitzka et al. 2012). A reduction at the C3 position would need to occur for ester formation. The C2 carboxylic acid methyl ester is the moiety responsible for the binding of cocaine onto the dopamine transporter (Carroll et al. 1992).

2.3 TA Modifications

The next step in tropane alkaloid biosynthesis is the reduction of the keto group found in both tropinone (Solanaceae) and methylecgonone (Erythroxylaceae). An enzyme specific for the reduction of tropinone to tropine was discovered and purified from the roots of *D. stramonium* and determined to require the cofactor NADPH (Koelen and Gross 1982). Another enzyme that also reduces tropinone but produces pseudotropine was purified from the roots of *H. niger* (Dräger et al. 1988). Concurrently, it was determined that pseudotropine does not spontaneously isomerize into tropine. These enzymes are known as tropinone reductase (TR) enzymes, for members of the Solanaceae family, and are a part of the short-chain dehydrogenase/reductase (SDR) family that catalyzes NAD(P)(H)-dependent monomeric oxidoreductase reactions whose activity controls the metabolic flux toward tropane alkaloid biosynthesis downstream (Kavanagh et al. 2008; Drager 2006). The SDR family of enzymes share a conserved active site with a catalytic residue motif YxxxK and have a common tertiary "Rossmann-fold" structure, a conserved motif that consists of two pairs of α-helices and six pairs of β-sheets, and a dinucleotide cofactor-binding motif (Kavanagh et al. 2008; Moummou et al. 2012). There are two distinct types of NADPH-dependent tropinone reductases in the Solanaceae family, tropinone reductase I (TRI, EC 1.1.1.206) and tropinone reductase II (TRII, EC 1.1.1.236). Many genes from different solanaceous species encoding TRI and TRII have been isolated since their initial discovery (Dräger and Schaal 1994; Kai et al. 2009; Keiner et al. 2002; Portsteffen et al. 1994; Nakajima et al. 1998). TRI and TRII share more than 50% amino acid sequence similarity and are presumed to have evolved from a common ancestor (Drager 2006). Additionally, a change of as small as five amino acids is required to change the stereospecificity of the reaction product (Nakajima et al. 1999). TRI catalyzes the reduction of the 3-keto functional group in tropinone into the 3α-hydroxyl configuration forming tropine, while TRII catalyzes a similar reduction but into the 3β-hydroxyl configuration forming pseudotropine which is then converted into calystegines. These two distinct tropinone reductases are attributed to a gene duplication event in the Solanaceae family (Nakajima et al. 1993). To the best of our knowledge, there is no evidence of interconversion between tropine and pseudotropine which suggests a branching of tropane alkaloid metabolism (Drager 2006; Yamada et al. 1990). In potato, these enzymes are localized in the tuber and roots based on immunolocalization experiments (Kaiser et al. 2006). Tropinone reductase was first reported in *D. stramonium* root cultures and reported to only catalyze the 3α-configuration in tropine (Koelen and Gross 1982).

Crystal structures of TRI and TRII from *D. stramonium* reveal that the overall fold between the two enzymes is almost identical (Nakajima et al. 1998). The two enzymes share 64% of the same amino acid residues and are thought to have diverged relatively recently from a common ancestral protein. These crystal structures confirm that the stereospecificity of TRI and TRII is due to the orientation of tropinone inside the enzyme. The selective stereochemistry of TRI is facilitated by a positive charge generated by a histidine residue that repels tropinone's positively charged nitrogen atom, therefore altering the orientation of tropinone within the active site. In TRII, this histidine residue is substituted with a tyrosine residue, and the active site of TRII is negatively charged due to a glutamate residue. The glutamate residue is substituted with a valine residue in TRI. This charge difference within the active site helps explain the stereospecificity of products between these two reductases and the binding orientation of tropinone. In *H. niger*, it was revealed that TRI is capable of catalyzing a reversible reaction; however, TRII can only catalyze the reaction in one direction (Hashimoto et al. 1992).

In 2008 Dräger et al. found a tropinone reductase-like gene in the transcriptome of *Cochlearia officinalis* of the Brassicaceae family, which also reduces tropinone. However, this reductase is capable of producing both tropine and pseudotropine in equal ratios using NADPH + H^+ as a co-substrate, in contrast to the stereospecific tropinone reductases found in the Solanaceae (Brock et al. 2008). This reductase (CoTR) is also more promiscuous than other tropinone reductases, accepting a broader range of substrates such as several synthetic ketones that had higher affinity and faster turnover than observed for tropinone. Moreover, CoTR is closely related to other members of the SDR family of enzymes in the Brassicaceae than it is to either tropinone reductases found in the Solanaceae, based on phylogenetic analysis.

Cocaine is one of the most well-known tropane alkaloid metabolites and can be found in multiple species within the Erythroxylaceae (Plowman and Rivier 1983). The biosynthesis of cocaine requires a similar reductase enzyme for the reduction of the 3-keto function of methylecgonone (2β-carbomethoxy-3-tropinone) into methylecgonine (2β-carbomethoxy-3β-tropine). Using a homology-based approach of tropinone reductase sequences from the Solanaceae, several tropinone reductase homologs were found within the *E. coca* transcriptome. Unfortunately, these candidates failed to yield reduction activity in methylecgonone when heterologously expressed in *E. coli* (Jirschitzka et al. 2012). Using crude extracts from coca leaves, Jirschitzka et al. were able to purify methylecgonone reductase (MecgoR, EC 1.1.1.334) and discovered that this enzyme was far different from the tropinone reductase enzymes responsible for the reduction of the 3-keto function in the

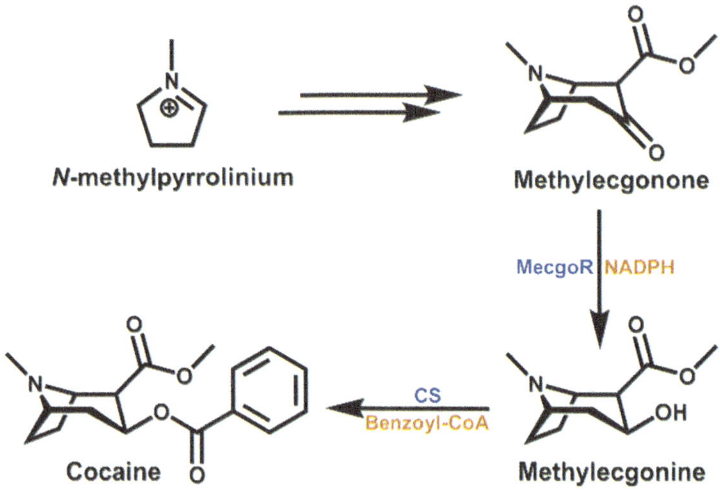

Scheme 2.6 Cocaine biosynthesis in *E. coca*. The enzymes depicted in blue are the methylecgonone reductase (MecgoR) and cocaine synthase (CS). The double arrows depict more than one enzymatic step is needed

Solanaceae (Scheme 2.6). When analyzed next to any tropinone reductases, MecgoR shares an overall identity of less than 10% at the amino acid level. It was found that MecgoR belongs to the aldo-keto reductase (AKR) superfamily whose enzymes are involved in a variety of metabolic pathways such as cocaine, codeine, and chalcone biosynthesis (Torres 2018; Barski et al. 2008). MecgoR shares 50–70% identity to the AKRs such as codeinone reductase (COR), chalcone reductase (CHR), and 1,2-dehydroreticuline reductase (DRR) despite using completely different substrates involved in different biochemical pathways (Dastmalchi et al. 2018). AKRs display a TIM-barrel motif containing an eight-strand β-barrel surrounded by α-helices with the cofactor-binding site located at the C-terminus of the β-barrel (Torres 2018; Penning 2015). AKRs use NADP or NADPH as a cofactor, and those that have been characterized so far all share the common α/β-barrel motif (Jez et al. 1997).

While the conversion of tropinone to pseudotropine by TRII in solanaceous plants is nonreversible, MecgoR found in *E. coca* is able to catalyze the reaction in both the reduction of methylecgonone and oxidation of methylecgonine. It should be noted that the oxidation of methylecgonine occurs only at its pH optimum of 9.8, which is unlikely to occur for a cytosolic enzyme (Jirschitzka et al. 2012). The preferred cofactor of MecgoR is NADPH but can successfully use NADH as a cofactor with only a 14% decrease in activity when compared to NADPH. MecgoR is a stereospecific enzyme responsible for the conversion of methylecgonone to the 3β-hydroxy-containing molecule methylecgonine. It can also use tropinone as a substrate but only yields pseudotropine exclusively. This stereospecificity corresponds with previous research regarding tropane alkaloid metabolites in the Erythroxylum family which predominantly consists of alkaloids with a β-configuration at the C3 position on the tropane core scaffold (Johnson 1995,

1996). Interestingly, the MecgoR protein is localized in the palisade parenchyma and has the highest activity in young developing leaves in contrast to the tropinone reductase proteins of the Solanaceae which are localized in the roots (Jirschitzka et al. 2012; Hashimoto et al. 1992; Nakajima and Hashimoto 1999). All of the evidence above provides strong indications that tropane alkaloid biosynthesis has independently evolved more than once in different plant lineages (Jirschitzka et al. 2012).

Although there are no characterized crystal structures of MecgoR, a homology model using codeinone reductase was made (Torres 2018). The model revealed an Ala25 residue predicted to be involved in the orientation and stabilization of the nicotinamide ring of NADPH in the cofactor-binding site. MecgoR and chalcone reductase also share this conserved Ala25residue. The NADPH orientation facilitates the hydride transfer from NADPH to the carbonyl group of the substrate. AKRs are known to catalyze a sequential bi-bi reaction mechanism in which the cofactor is the first to bind to the enzyme and last to leave. The reaction mechanism of AKRs shows that the pro-(*R*) hydride is transferred from NADPH instead of the pro-(*S*) hydride which is observed in the mechanism of tropinone reductase and other SDRs due to the orientation difference of NADPH relative to the substrate. Further modeling or crystallization of the MecgoR enzyme is required for further structure-function analysis.

Methylecgonine has little physiological activity until it is converted to the benzoic ester of methylecgonine, cocaine (Williams et al. 1977). Based on feeding studies using *trans*-[3-^{13}C,^{14}C]-cinnamic acid and the *N*-acetylcysteamine thioester of [3-^{13}C,^{14}C]-*trans*-cinnamic acid, it was hypothesized that an acyltransferase in *E. coca* employs benzoyl-CoA as the activated acid (Bjorklund and Leete 1992). Leete et al. first proposed that the esterification of tropane alkaloids in *E. coca* would utilize CoA-activated thioesters (Leete et al. 1988b). It was predicted that methylecgonine undergoes esterification with a benzoyl moiety that uses benzoyl-CoA as the activated acyl donor. However, it was unclear whether it was derived from benzoyl-CoA or benzaldehyde, but it is clear that the moiety was found to originate from cinnamic acid (Bjorklund and Leete 1992; Leete et al. 1988b). The acylation of secondary metabolites in plants has been depicted in three families of acyltransferases, but only BAHD acyltransferases have been reported to use activated acyl-CoA thioesters (D'Auria 2006). Based on reported properties for the tigloyl-CoA:pseudotropine acyltransferase in *D. stramonium* and the CoA-dependent nature of this enzyme, it was hypothesized that the enzyme involved in facilitating this type of reaction is a member of the BAHD acyltransferase superfamily who are well-known to engage in secondary metabolite modifications of amides and esters (Rabot et al. 1995). Schmidt et al. identified a BAHD enzyme from *E. coca* capable of catalyzing the acylation of 3β-hydroxyl group of methylecgonine using a benzoyl-CoA thioester as the acyl donor which was named cocaine synthase (EcCS), the last step in cocaine biosynthesis (Schmidt et al. 2015). This enzyme can use benzoyl-CoA and cinnamoyl-CoA thioesters to produce cocaine and cinnamoylcocaine, respectively. Ester formation using pseudotropine (3β-OH, no C2 carbomethoxy function) and benzoyl-CoA as a substrate for EcCS is possible, retaining 80% of its activity when compared to methylecgonine and benzoyl-CoA. However, ester formation was not achieved when using tropine (3α-OH) or nortropine (3α-OH, no *N*-methyl group) as substrates in this report. Interestingly, a recent report from Srinivasan et al. demonstrated that EcCS can catalyze the formation of cinnamoyltropine using tropine and cinnamoyl-CoA as substrates (Srinivasan and Smolke 2019). This report contradicts previous experiments that demonstrated substrate specificity of EcCS for the 3β-hydroxyl function of methylecgonine and pseudotropine rather than the 3α-hydroxyl function of tropine for ester formation (Schmidt et al. 2015). This unusual promiscuity of cocaine synthase can be used to produce novel tropane alkaloid derivatives for both medical and biotechnological applications.

The biosynthesis and accumulation of tropane alkaloids in *E. coca* occur in the same tissue. Immunolocalization experiments show the accumulation of EcCS in the palisade layer and in the spongy mesophyll. This is opposed to solanaceous plants where the biosynthetic pathway of tropane alkaloids occurs in the roots while metabolites are accumulated in the shoots or aboveground portion (Ziegler and Facchini 2008).

Currently there are no crystal structure or modeling data on EcCS; however, there is data on the BAHD superfamily of enzymes. The first BAHD crystal structure was that of a vinorine synthase and can be used to help model other novel BAHD acyltransferases (Ma et al. 2005). EcCS, like other characterized BAHDs, is monomeric in structure and a member of clade III of the BAHD superfamily who can use a wide variety of alcohol substrates with CoA being the major acyl donor in most of these reactions (D'Auria 2006; Schmidt et al. 2015). There are two conserved motifs that are present in all characterized BAHD acyltransferases. The first motif is the DFGWG motif which is located on the C-terminus and appears to have no role in the catalytic activity of the enzyme but appears to be a structural domain involved in maintaining the CoA binding pocket for BAHD acyltransferases (Ma et al. 2005). The second motif is the HXXXD motif and is important for the catalytic activity of BAHD acyltransferases. The histidine residue on the HXXXD motif acts as a general base to deprotonate a nitrogen or oxygen atom on a substrate facilitating a nucleophilic attack on the carbonyl atom of the CoA thioester forming a tetrahedral intermediate. This intermediate then becomes protonated again producing a free CoA and an acylated substrate.

The hydroxyl group rearrangement of the phenyllactic acid moiety of littorine in tropane alkaloid side chain biosynthesis is poorly understood and a point of interest. Littorine is an important intermediate in solanaceous plants, occurring in atropine and scopolamine biosynthesis (Hashimoto et al. 1993). There had been discussion regarding the acylation of secondary metabolites that can be catalyzed by serine carboxypeptidase-like (SCPL) acyltransferases which use 1-*O*-β-glucose esters as the acyl donor instead of activated CoA thioesters (Bontpart et al. 2015). It was postulated that an enzyme analogous to cocaine synthase was utilizing an activated phenyllactoyl-CoA thioester for littorine biosynthesis, but a recent report casts doubts on this hypothesis (Robins et al. 1994a; Qiu et al. 2019; Kohnen-Johannsen and Kayser 2019). The new study on littorine biosynthesis discovered that a phenyllactate UDP-glycosyltransferase (UGT1) and littorine synthase (LS) are involved in tropane alkaloid biosynthesis in *A. belladonna (*Qiu et al. 2019*)*. UGT1 is responsible for the conversion of phenyllactate into phenyllactylglucose that is subsequently used as a substrate by LS. Suppression of either gene showed an increase in tropine levels and a subsequent decrease in the levels of hyoscyamine and scopolamine in *A. belladonna* root cultures. qPCR analysis of both genes shows high expression levels in the secondary roots with lower levels detected in the primary roots. Functional identification of UGT1 and LS genes was mediated through agrobacterium transformation coupled with tropine feeding. The production and accumulation of littorine in *N. benthamiana*, a plant species that does not produce any tropane alkaloids, was detected. Littorine has been detected in the roots of *D. myoporoides* through MALDI-MSI, and quantification in roots was done using HPLC-MS to reveal constant levels of littorine, regardless of growth stage of the plant (Kohnen et al. 2018).

Early feeding studies of radiolabeled compounds have attempted the elucidation of the hyoscyamine (Leete et al. 1975; Ansarin and Woolley 1994; Robins et al. 1994b, 1995; Chesters et al. 1995). The leading hypothesis for the last steps of hyoscyamine biosynthesis involved a cytochrome p450 coupled with an alcohol dehydrogenase, based on quantum chemistry calculations and feeding studies (Sandala et al. 2008). The discovery of the cytochrome p450 oxidoreductase, littorine mutase/monooxygenase (CYP80F1, EC 1.6.2.4), demonstrated that the 5′-deoxyadenosyl radical reaction mechanism was unlikely to play a direct role in the conversion of littorine to hyoscyamine (Robins et al. 1994b, 1995). The conversion of littorine into hyoscyamine is a two-step process in which (*R*)-littorine must first undergo an oxidation-reduction reaction catalyzed by littorine mutase/ monooxygenase (CYP80F1) to yield the intermediate hyoscyamine aldehyde. The second part of this reaction is thought to involve an NADPH-dependent alcohol dehydrogenase to finish the conversion of hyoscyamine aldehyde to (*S*)-hyoscyamine, but this enzyme has yet to be identified. Virus-induced gene silencing (VIGS) techniques were used to suppress the expression of CYP80F1 that resulted in the lower levels of hyoscyamine and promoted littorine accumulation (Li et al. 2006). Additionally, CYP80F1 expression in tobacco roots supplemented with (*R*)-littorine resulted in hyoscyamine accumulation further supporting the role of CYP80F1 in the conversion of litterine to hyoscyamine. RT-PCR revealed that *CYP80F1* is only expressed in the root tissues of *H. niger*. Through the use of arylfluorinated analogs of (*R*)-littorine (natural isomer) and (*S*)-littorine (unnatural isomer) as substrates for CYP80F1, it was determined that the enzyme-catalyzed hydroxylation occurs via a benzylic carbocation intermediate (Nasomjai et al. 2009).

The formation of scopolamine from hyoscyamine is carried out by hyoscyamine-6β-hydroxylase (H6H; EC 1.14.11.11 and EC 1.14.20.13). The purified H6H enzyme from *H. niger* was demonstrated to be a 2-oxoglutarate-dependent dioxygenase, a bifunctional dioxygenase containing both hydroxylase and epoxidase activity (Hashimoto et al. 1993; Hashimoto and Yamada 1986). Past studies have suggested that the amount of H6H enzyme in plants may be rate-limiting for the accumulation of scopolamine (Hashimoto et al. 1993). Fe^{2+} and L-ascorbate are essential cofactors for the dioxygenase activity of H6H (Averesch and Kayser 2014; Hashimoto et al. 1991). Hyoscyamine 6β-hydroxylase contains two iron binding motifs as well as an α-ketoglutarate binding motif (Fischer et al. 2018). The H6H enzyme first catalyzes the 6β-hydroxylation of *L*-hyoscyamine's tropane ring into anisodamine (6β-hydroxyhyoscyamine) (Averesch and Kayser 2014). An intramolecular epoxidation formation then takes place through the removal of the 6-β-hydrogen that directly results in scopolamine. Even though both reactions are catalyzed by the same enzyme, different EC numbers are given for each step. The H6H enzyme has been discovered in several members of the Solanaceae including *A. belladonna*, *H. niger*, and *A. acutangulus*.

Localization of H6H was found exclusively in the pericycle of roots (Hashimoto et al. 1991). Similar to other solanaceous molecules, scopolamine is primarily synthesized in the roots and then stored in the leaves (Hashimoto and Yamada 1986; Hashimoto et al. 1991; Zhao et al. 2017; Suzuki et al. 1999). The exact transport mechanisms of how scopolamine and other tropane alkaloids are shuttled from the roots to the leaves are not fully understood. One proposal suggests scopolamine glucoside is involved in the transport of scopolamine to the leaves. However, due to the low abundance of scopolamine glucoside found, this hypothesis was excluded from being the specific transport system of scopolamine. Simple diffusion of tropane alkaloids through the xylem is deemed unlikely due to the fact that tropane alkaloids are found in higher concentrations in the leaf blades rather than in the vascular tissue (Kohnen et al. 2018). An intriguing observation of the H6H enzyme is its localization in the pericycle, a location absent of TRI (Hashimoto et al. 1991). It has been noted that endodermal cells contain a Casparian strip that blocks metabolite transport in plants, suggesting only tropane intermediates produced on the inner side of root endodermal cells can participate in scopolamine biosynthesis (Nakajima and Hashimoto 1999; Esau 1977).

Stereospecificity studies of H6H reveal that C7 hydroxylation of hyoscyamine is possible but not as favorable as C6 hydroxylation (Ushimaru et al. 2018, 2019). Recently, a SUMO-tagged BsH6H from *Brugmansia sanguinea* was reported to have approximately ten times greater epoxidase activity than any other previously characterized H6H homologs. The 3D structure of BsH6H was modeled to the closest PDB hit of *A. thaliana* anthocyanidin synthase which only shares a 28% sequence identity with BsH6H (Fischer et al. 2018). Comparisons with other H6H homologs and structural overlay reveal an overall Rossmann fold with a flexible loop between Lys91-Asp130. Additionally, 30 amino acids of the N-terminal region of all H6H homologs suggest high flexibility with no clear secondary structural elements. Therefore, when a truncated version of H6H from *A. belladonna* with 30 amino acids missing on the N-terminus was made, there was no significant change in enzyme activity, and it was predicted to have a higher crystallization potential. This potential could be useful for ascertaining a crystal structure of H6H to which there are currently none.

2.4 Conclusions

Research on tropane alkaloid biosynthesis in the last decade began with a plethora of open questions regarding enzymes and substrates and has, in the interim, begun to fill in those gaps. It is clear that the entire pathway, at least for the main solanaceous species, is nearly completed. As with all scientific endeavors, the discovery and initial characterization of the structure-function relationships of these enzymes only reveals more questions and a need for continuing tropane research into the future. Increasing amounts of evidence suggest that the tropane biosynthetic pathway is polyphyletic. This then begs the question, what biological pressures were there in the different lineages that yielded a common structure? What are the real biological functions of tropanes in an ecological context? Clearly, several studies already exist, suggesting that tropanes serve as defensive compounds against herbivores (Shonle and Bergelson 2000; Wilson et al. 2018). However, there are many open questions regarding the diversity of structures within the family and the interactions they are mediating.

Of the enzymes identified and characterized to date, many structure-function relationships remain unclear. It is now a task for structural studies to elucidate dynamics of folding and the identification of residues important not only for catalysis but for regulation as well. It is interesting that at least half of the enzymes in the tropane biosynthetic pathway exist as dimers. New studies will be necessary to identify allosteric regulators of these enzymes, and this in turn will give rise to the need for flux analysis and experiments involving cross talk and inhibition. Lastly, these subjects also relate to storage and turnover of tropanes within the plant, a key area of study in which there is little to no data available.

The lesser studied families containing members with the ability to produce tropanes are now ripe for discovery of even more novel enzymes. With the increase in sequencing capabilities and the accompanying reduction in costs, more genetic resources will be available for gene discovery studies in families such as the Brassicaceae, Convolvulaceae, and Proteaceae. Regardless if new enzymes are identified in the core metabolic pathway, it is inevitable that the diversity of enzymes involved in decorating the core structures will be discovered. Alongside these discoveries, the fields of synthetic biology and computational enzyme design will benefit and assist in addressing issues regarding the production of pharmaceuticals and bioactive tropanes for pharmaceutical purposes in a "green chemistry" context. It is clear that the future of tropane research moving forward into the next decade will focus on novel tropanes and metabolic engineering utilizing our new gains in structure-function relationships.

References

Abraham TW, Leete E (1995) New intermediate in the biosynthesis of the tropane alkaloids in Datura-innoxia. J Am Chem Soc 117(31):8100–8105

Ahou A et al (2014) A plant spermine oxidase/dehydrogenase regulated by the proteasome and polyamines. J Exp Bot 65(6):1585–1603

Ansarin M, Woolley JG (1994) The biosynthesis of tropic acid. 5. The rearrangement of phenyllactate in the biosynthesis of tropic acid. Phytochemistry 35(4):935–939

Austin MB, Noel JP (2003) The chalcone synthase superfamily of type III polyketide synthases. Nat Prod Rep 20(1):79–110

Averesch NJH, Kayser O (2014) Assessing heterologous expression of hyoscyamine 6-β-hydroxylase – a feasibility study. Proc Chem 13:69–78

Barski OA, Tipparaju SM, Bhatnagar A (2008) The aldo-keto reductase superfamily and its role in drug metabolism and detoxification. Drug Metab Rev 40(4):553–624

Bedewitz MA et al (2018) Tropinone synthesis via an atypical polyketide synthase and P450-mediated cyclization. Nat Commun 9(1):5281

Biastoff S, Brandt W, Drager B (2009a) Putrescine N-methyltransferase--the start for alkaloids. Phytochemistry 70(15-16):1708–1718

Biastoff S et al (2009b) Evolution of putrescine N-methyltransferase from spermidine synthase demanded alterations in substrate binding. FEBS Lett 583(20):3367–3374

Bjorklund JA, Leete E (1992) Biosynthesis of the tropane alkaloids and related-compounds. 49. Biosynthesis of the benzoyl moiety of cocaine from cinnamic acid via (R)-(+)-3-hydroxy-3-phenylpropanoic acid. Phytochemistry 31(11):3883–3887

Bontpart T et al (2015) BAHD or SCPL acyltransferase? What a dilemma for acylation in the world of plant phenolic compounds. New Phytol 208(3):695–707

Borrell A et al (1995) Arginine decarboxylase is localized in chloroplasts. Plant Physiol 109 (3):771–776

Bortolotti C et al (2004) Localization of arginine decarboxylase in tobacco plants. Physiol Plant 120 (1):84–92

Brock A, Brandt W, Drager B (2008) The functional divergence of short-chain dehydrogenases involved in tropinone reduction. Plant J 54(3):388–401

Carroll FI et al (1992) Probes for the cocaine receptor. Potentially irreversible ligands for the dopamine transporter. J Med Chem 35(10):1813–1817

Cervelli M et al (2001) A barley polyamine oxidase isoform with distinct structural features and subcellular localization. Eur J Biochem 268(13):3816–3830

Chattopadhayay MK et al (2002) Protective role of exogenous polyamines on salinity-stressed rice (Oryza sativa) plants. Physiol Plant 116(2):192–199

Chattopadhyay MK, Tabor CW, Tabor H (2003) Polyamines protect Escherichia coli cells from the toxic effect of oxygen. Proc Natl Acad Sci U S A 100(5):2261–2265

Chesters NCJE, Ohagan D, Robins RJ (1995) The biosynthesis of tropic acid - the (R)-D-phenyllactyl moiety is processed by the mutase involved in hyoscyamine biosynthesis in Datura-stramonium. J Chem Soc 2:127–128

Coleman CS, Stanley BA, Pegg AE (1993) Effect of mutations at active-site residues on the activity of ornithine decarboxylase and its inhibition by active-site-directed irreversible inhibitors. J Biol Chem 268(33):24572–24579

Cona A et al (2006) Functions of amine oxidases in plant development and defence. Trends Plant Sci 11(2):80–88

Dang TT, Facchini PJ (2014) CYP82Y1 is N-methylcanadine 1-hydroxylase, a key noscapine biosynthetic enzyme in opium poppy. J Biol Chem 289(4):2013–2026

Dastmalchi M et al (2018) Codeinone reductase isoforms with differential stability, efficiency and product selectivity in opium poppy. Plant J. https://doi.org/10.1111/tpj.13975

D'Auria JC (2006) Acyltransferases in plants: a good time to be BAHD. Curr Opin Plant Biol 9 (3):331–340

Delauney AJ, Verma DPS (1993) Proline biosynthesis and osmoregulation in plants. Plant J 4 (2):215–223

Dillehay TD et al (2010) Early Holocene coca chewing in northern Peru. Antiquity 84 (326):939–953

Docimo T et al (2012) The first step in the biosynthesis of cocaine in Erythroxylum coca: the characterization of arginine and ornithine decarboxylases. Plant Mol Biol 78(6):599–615

Drager B (2004) Chemistry and biology of calystegines. Nat Prod Rep 21(2):211–223

Drager B (2006) Tropinone reductases, enzymes at the branch point of tropane alkaloid metabolism. Phytochemistry 67(4):327–337

Dräger B, Schaal A (1994) Tropinone reduction in Atropa belladonna root cultures. Phytochemistry 35(6):1441–1447

Dräger B, Hashimoto T, Yamada Y (1988) Purification and characterization of pseudotropine forming tropinone reductase from Hyoscyamus niger root cultures. Agric Biol Chem 52 (10):2663–2667

Endo T et al (1988) Non-enzymatic synthesis of hygrine from acetoacetic acid and from acetonedicarboxylic acid. FEBS Lett 234(1):86–90

Esau K (1977) Anatomy of the seed plants, 2nd edn. Wiley, New York, p 576

Feth F, Wray V, Wagner KG (1985) Determination of methylputrescine oxidase by high performance liquid chromatography. Phytochemistry 24(8):1653–1655

Fincato P et al (2011) Functional diversity inside the Arabidopsis polyamine oxidase gene family. J Exp Bot 62(3):1155–1168

Fischer C et al (2018) Isolation, expression and biochemical characterization of recombinant hyoscyamine-6beta-hydroxylase from Brugmansia sanguinea - tuning the scopolamine production. MedChemComm 9(5):888–892

Geiger PL, Hesse H (1833) Darstellung des atropins. Ann Pharm 5(1):43–81

Hartzoulakis B et al (2007) Discovery of inhibitors of the pentein superfamily protein dimethylarginine dimethylaminohydrolase (DDAH), by virtual screening and hit analysis. Bioorg Med Chem Lett 17(14):3953–3956

Hashimoto T, Yamada Y (1986) Hyoscyamine 6beta-hydroxylase, a 2-oxoglutarate-dependent dioxygenase, in alkaloid-producing root cultures. Plant Physiol 81(2):619–625

Hashimoto T, Mitani A, Yamada Y (1990) Diamine oxidase from cultured roots of Hyoscyamus niger: its function in tropane alkaloid biosynthesis. Plant Physiol 93(1):216–221

Hashimoto T et al (1991) Hyoscyamine 6β-hydroxylase, an enzyme involved in tropane alkaloid biosynthesis, is localized at the pericycle of the root. J Biol Chem 266(7):4648–4653

Hashimoto T et al (1992) Two tropinone reductases with distinct stereospecificities from cultured roots of Hyoscyamus niger. Plant Physiol 100(2):836–845

Hashimoto T, Matsuda J, Yamada Y (1993) Two-step epoxidation of hyoscyamine to scopolamine is catalyzed by bifunctional hyoscyamine 6β-hydroxylase. FEBS Lett 329(1-2):35–39

Hashimoto T et al (1998) Intraspecific variability of the tandem repeats in Nicotiana putrescine N-methyltransferases. Plant Mol Biol 37(1):25–37

Heim WG et al (2007) Cloning and characterization of a Nicotiana tabacum methylputrescine oxidase transcript. Phytochemistry 68(4):454–463

Hoye TR et al (2000) N-methylputrescine oxidation during cocaine biosynthesis: study of prochiral methylene hydrogen discrimination using the remote isotope method. Org Lett 2(1):3–5

Huang JP et al (2019) Tropane alkaloids biosynthesis involves an unusual type III polyketide synthase and non-enzymatic condensation. Nat Commun 10(1):4036

Humphrey AJ, O'Hagan D (2001) Tropane alkaloid biosynthesis. A century old problem unresolved. Nat Prod Rep 18(5):494–502

Imanishi S et al (1998) Differential induction by methyl jasmonate of genes encoding ornithine decarboxylase and other enzymes involved in nicotine biosynthesis in tobacco cell cultures. Plant Mol Biol 38(6):1101–1111

Janowitz T, Kneifel H, Piotrowski M (2003) Identification and characterization of plant agmatine iminohydrolase, the last missing link in polyamine biosynthesis of plants. FEBS Lett 544 (1-3):258–261

Jez JM et al (1997) Comparative anatomy of the aldo-keto reductase superfamily. Biochem J 326 (Pt 3):625–636

Jirschitzka J et al (2012) Plant tropane alkaloid biosynthesis evolved independently in the Solanaceae and Erythroxylaceae. Proc Natl Acad Sci U S A 109(26):10304–10309

Johnson EL (1995) Content and distribution of erythroxylum-coca leaf alkaloids. Ann Bot 76 (4):331–335

Johnson EL (1996) Alkaloid content in Erythroxylum coca tissue during reproductive development. Phytochemistry 42(1):35–38

Junker A et al (2013) Evolution of the key alkaloid enzyme putrescine N-methyltransferase from spermidine synthase. Front Plant Sci 4:260

Kacprzak KM (2013) Chemistry and biology of cinchona alkaloids. In: Ramawat KG, Mérillon J-M (eds) Natural products: phytochemistry, botany and metabolism of alkaloids, phenolics and terpenes. Springer, Berlin, pp 605–641

Kaczkowski J, Mothes K, Schutte HR (1961) Die rolle des acetats in der biosynthese der tropanalkaloide. Biochim Biophys Acta 46(3):588

Kai G et al (2009) Molecular cloning and characterization of two tropinone reductases in Anisodus acutangulus and enhancement of tropane alkaloid production in AaTRI-transformed hairy roots. Biotechnol Appl Biochem 54(3):177–186

Kaiser H et al (2006) Immunolocalisation of two tropinone reductases in potato (Solanum tuberosum L.) root, stolon, and tuber sprouts. Planta 225(1):127–137

Kamada-Nobusada T et al (2008) A putative peroxisomal polyamine oxidase, AtPAO4, is involved in polyamine catabolism in Arabidopsis thaliana. Plant Cell Physiol 49(9):1272–1282

Katoh A, Shoji T, Hashimoto T (2007) Molecular cloning of N-methylputrescine oxidase from tobacco. Plant Cell Physiol 48(3):550–554

Kavanagh KL et al (2008) Medium- and short-chain dehydrogenase/reductase gene and protein families : the SDR superfamily: functional and structural diversity within a family of metabolic and regulatory enzymes. Cell Mol Life Sci 65(24):3895–3906

Keiner R et al (2002) Molecular cloning, expression and characterization of tropinone reductase II, an enzyme of the SDR family in Solanum tuberosum (L.). Plant Mol Biol 48(3):299–308

Kim DW et al (2014) Polyamine oxidase5 regulates arabidopsis growth through thermospermine oxidase activity. Plant Physiol 165(4):1575–1590

Kim N et al (2016) Tropane and granatane alkaloid biosynthesis: a systematic analysis. Molecules 21:11

Koelen KJ, Gross GG (1982) Partial purification and properties of tropine dehydrogenase from root cultures of Datura stramonium. Planta Med 44(4):227–230

Kohnen KL et al (2018) Localization and organization of scopolamine biosynthesis in duboisia myoporoides R. Br. Plant Cell Physiol 59(1):107–118

Kohnen-Johannsen KL, Kayser O (2019) Tropane alkaloids: chemistry, pharmacology, biosynthesis and production. Molecules 24:4

Kusano T et al (2008) Polyamines: essential factors for growth and survival. Planta 228(3):367–381

Lee YS, Cho YD (2001) Identification of essential active-site residues in ornithine decarboxylase of Nicotiana glutinosa decarboxylating both L-ornithine and L-lysine. Biochem J 360(3):657–665

Leete E (1962) Stereospecific incorporation of ornithine into tropine moiety of hyoscyamine. J Am Chem Soc 84(1):55–57

Leete E (1982) Biosynthesis of the pyrrolidine rings of cocaine and cuscohygrine from [5-14C]-ornithine via a symmetrical intermediate. J Am Chem Soc 104(5):1403–1408

Leete E (1983) Chemistry of the tropane alkaloids. 31. Biosynthesis of cocaine and cuscohygrine from [1-^{14}C]-labeled acetate and [4-^{3}H]-labeled phenylalanine in erythroxylon-coca. Phytochemistry 22(3):699–704

Leete E (1985) Spermidine: an indirect precursor of the pyrrolidine rings of nicotine and nornicotine in Nicotiana glutinosa. Phytochemistry 24(5):957–960

Leete E (1990) Recent developments in the biosynthesis of the tropane alkaloids. Planta Med 56 (4):339–352

Leete E, Marion L, Spenser ID (1954) The biogenesis of alkaloids. 12. The mode of formation of the tropine base of hyoscyamine. Can J Chem 32(12):1116–1123

Leete E, Kowanko N, Newmark RA (1975) Use of carbon-13 nuclear magnetic resonance to establish that the biosynthesis of tropic acid involves an intramolecular rearrangement of phenylalanine. J Am Chem Soc 97(23):6826–6830

Leete E, Kim SH, Rana J (1988a) Chemistry of the tropane alkaloids and related-compounds. 38. The incorporation of [2-^{13}C,^{14}C,^{15}N]-1-methyl-D^{1}-pyrrolinium chloride into cuscohygrine in Erythroxylum coca. Phytochemistry 27(2):401–406

Leete E, Bjorklund JA, Kim SH (1988b) Chemistry of the tropane alkaloids and related-compounds. 40. The biosynthesis of the benzoyl moiety of cocaine. Phytochemistry 27 (8):2553–2556

Leete E et al (1991) Late intermediates in the biosynthesis of cocaine-4-(1-methyl-2-pyrrolidinyl)-3-oxobutanoate and methyl ecgonine. J Am Chem Soc 113(24):9286–9292

Lewis RS et al (2010) Three nicotine demethylase genes mediate nornicotine biosynthesis in Nicotiana tabacum L.: functional characterization of the CYP82E10 gene. Phytochemistry 71 (17-18):1988–1998

Li R et al (2006) Functional genomic analysis of alkaloid biosynthesis in Hyoscyamus niger reveals a cytochrome P450 involved in littorine rearrangement. Chem Biol 13(5):513–520

Liebisch HW, Schütte HR (1967) Zur Biosynthese der Tropanalkaloide. VIII. Vorstufen des Pyrrolidinringes. Z Pflanzenphysiol 57(5):434–439

Liebisch HW et al (1972) Zur biosynthese der tropanalkaloide. XI. Die bildung der C3-brücke des tropins. Z Pflanzenphysiol 67(1):1–9

Liu T et al (2014a) Oryza sativa polyamine oxidase 1 back-converts tetraamines, spermine and thermospermine, to spermidine. Plant Cell Rep 33(1):143–151

Liu T et al (2014b) Polyamine oxidase 7 is a terminal catabolism-type enzyme in Oryza sativa and is specifically expressed in anthers. Plant Cell Physiol 55(6):1110–1122

Lounasmaa M, Tamminen T (1993) The tropane alkaloids. In: Cordell GA (ed) The alkaloids. Academic, New York, pp 1–114

Ma X et al (2005) Crystal structure of vinorine synthase, the first representative of the BAHD superfamily. J Biol Chem 280(14):13576–13583

Malmberg RL et al (1992) Arginine decarboxylase of oats is clipped from a precursor into two polypeptides found in the soluble enzyme. Plant Physiol 100(1):146–152

Malmberg RL et al (1998) Molecular genetic analyses of plant polyamines. Crit Rev Plant Sci 17 (2):199–224

Matsuzaki R et al (1994) Generation of the topa quinone cofactor in bacterial monoamine oxidase by cupric ion-dependent autooxidation of a specific tyrosyl residue. FEBS Lett 351(3):360–364

Mein K (1833) Darstellung des Atropins in weissen Krystallen. Ann Pharmacother 6(1):67–72

Michael AJ et al (1996) Molecular cloning and functional identification of a plant ornithine decarboxylase cDNA. Biochem J 314(1):241–248

Miller-Fleming L et al (2015) Remaining mysteries of molecular biology: the role of polyamines in the cell. J Mol Biol 427(21):3389–3406

Minguet EG et al (2008) Evolutionary diversification in polyamine biosynthesis. Mol Biol Evol 25 (10):2119–2128

Mizusaki S et al (1973) Phytochemical studies on tobacco alkaloids. 16. Changes in activities of ornithine decarboxylase, putrescine N-methyltransferase and N-methyl-putrescine oxidase in tobacco roots in relation to nicotine biosynthesis. Plant Cell Physiol 14(1):103–110

Mo H et al (2015) Cotton polyamine oxidase is required for spermine and camalexin signalling in the defence response to Verticillium dahliae. Plant J 83(6):962–975

Mohan Chaudhuri M, Ghosh B (1985) Agmatine deiminase in rice seedlings. Phytochemistry 24 (10):2433–2435

Moschou PN et al (2008) Bridging the gap between plant and mammalian polyamine catabolism: a novel peroxisomal polyamine oxidase responsible for a full back-conversion pathway in Arabidopsis. Plant Physiol 147(4):1845–1857

Moummou H et al (2012) The plant short-chain dehydrogenase (SDR) superfamily: genome-wide inventory and diversification patterns. BMC Plant Biol 12:219

Naconsie M et al (2014) Molecular evolution of N-methylputrescine oxidase in tobacco. Plant Cell Physiol 55(2):436–444
Nakajima K, Hashimoto T (1999) Two tropinone reductases, that catalyze opposite stereospecific reductions in tropane alkaloid biosynthesis, are localized in plant root with different cell-specific patterns. Plant Cell Physiol 40(11):1099–1107
Nakajima K, Hashimoto T, Yamada Y (1993) Two tropinone reductases with different stereospecificities are short-chain dehydrogenases evolved from a common ancestor. Proc Natl Acad Sci U S A 90(20):9591–9595
Nakajima K et al (1998) Crystal structures of two tropinone reductases: different reaction stereospecificities in the same protein fold. Proc Natl Acad Sci U S A 95(9):4876–4881
Nakajima K et al (1999) Site-directed mutagenesis of putative substrate-binding residues reveals a mechanism controlling the different stereospecificities of two tropinone reductases. J Biol Chem 274(23):16563–16568
Nam KH, Lee SH, Lee J (1997) Purification and characterization of arginine decarboxylase from soybean (Glycine max) hypocotyls. Plant Cell Physiol 38(10):1150–1155
Nasomjai P et al (2009) Mechanistic insights into the cytochrome P450-mediated oxidation and rearrangement of littorine in tropane alkaloid biosynthesis. Chembiochem 10(14):2382–2393
Nocquet PA, Opatz T (2016) Total synthesis of (+/-)-scopolamine: challenges of the tropane ring. Eur J Org Chem 2016(6):1156–1164
Penning TM (2015) The aldo-keto reductases (AKRs): overview. Chem Biol Interact 234:236–246
Piotrowski M, Janowitz T, Kneifel H (2003) Plant C-N hydrolases and the identification of a plant N-carbamoylputrescine amidohydrolase involved in polyamine biosynthesis. J Biol Chem 278 (3):1708–1712
Plowman T, Rivier L (1983) Cocaine and cinnamoylcocaine content of Erythroxylum species. Ann Bot 51(5):641–659
Portsteffen A, Drager B, Nahrstedt A (1994) The reduction of tropinone in Datura stramonium root cultures by two specific reductases. Phytochemistry 37(2):391–400
Qiang W et al (2016) Functional characterisation of a tropine-forming reductase gene from Brugmansia arborea, a woody plant species producing tropane alkaloids. Phytochemistry 127:12–22
Qiu F et al (2019) Functional genomics analysis reveals two novel genes required for littorine biosynthesis. New Phytol 225:1906–1914
Rabot S, Peerless ACJ, Robins RJ (1995) Tigloyl-coa - pseudotropine acyl transferase - an enzyme of tropane alkaloid biosynthesis. Phytochemistry 39(2):315–322
Robins R et al (1994a) Phenyllactic acid but not tropic acid is an intermediate in the biosynthesis of tropane alkaloids in Datura and Brugmansia transformed root cultures. Planta 194(1):86–94
Robins RJ, Bachmann P, Woolley JG (1994b) Biosynthesis of hyoscyamine involves an intramolecular rearrangement of littorine. J Chem Soc Trans 1(6):615–619
Robins RJ et al (1995) The biosynthesis of hyoscyamine - the process by which Littorine rearranges to hyoscyamine. J Chem Soc Transact 1(4):481–485
Robins RJ et al (1997) The biosynthesis of tropane alkaloids in Datura stramonium: the identity of the intermediates between N-methylpyrrolinium salt and tropinone. J Am Chem Soc 119 (45):10929–10934
Romek KM et al (2016) Non-statistical 13C fractionation distinguishes co-incident and divergent steps in the biosynthesis of the alkaloids nicotine and tropine. J Biol Chem 291 (32):16620–16629
Sandala GM, Smith DM, Radom L (2008) The carbon-skeleton rearrangement in tropane alkaloid biosynthesis. J Am Chem Soc 130(32):10684–10690
Sandmeier E, Hale TI, Christen P (1994) Multiple evolutionary origin of pyridoxal-5'-phosphate-dependent amino acid decarboxylases. Eur J Biochem 221(3):997–1002
Schipper RG et al (2004) Intracellular localization of ornithine decarboxylase and its regulatory protein, antizyme-1. J Histochem Cytochem 52(10):1259–1266

Schmeller T et al (1995) Binding of tropane alkaloids to nicotinic and muscarinic acetylcholine receptors. Pharmazie 50(7):493–495

Schmidt GW et al (2015) The last step in cocaine biosynthesis is catalyzed by a BAHD acyltransferase. Plant Physiol 167(1):89–101

Šebela M et al (2001) FAD-containing polyamine oxidases: a timely challenge for researchers in biochemistry and physiology of plants. Plant Sci 160(2):197–207

Sekula B, Dauter Z (2019a) Structural study of agmatine iminohydrolase from medicago truncatula, the second enzyme of the agmatine route of putrescine biosynthesis in plants. Front Plant Sci 10:320

Sekula B, Dauter Z (2019b) Spermidine Synthase (SPDS) undergoes concerted structural rearrangements upon ligand binding - a case study of the two SPDS isoforms from Arabidopsis thaliana. Front Plant Sci 10:555

Sekula B et al (2016) Structural investigations of N-carbamoylputrescine amidohydrolase from medicago truncatula: insights into the ultimate step of putrescine biosynthesis in plants. Front Plant Sci 7:350

Shakeran Z et al (2015) Improvement of atropine production by different biotic and abiotic elicitors in hairy root cultures of Datura metal. Turk J Biol 39(1):111–118

Shonle I, Bergelson J (2000) Evolutionary ecology of the tropane alkaloids of Datura stramonium L.(Solanaceae). Evolution 54(3):778–788

Siminszky B et al (2005) Conversion of nicotine to nornicotine in Nicotiana tabacum is mediated by CYP82E4, a cytochrome P450 monooxygenase. Proc Natl Acad Sci U S A 102 (41):14919–14924

Slocum RD, Kaur-Sawhney R, Galston AW (1984) The physiology and biochemistry of polyamines in plants. Arch Biochem Biophys 235(2):283–303

Soeren JH (1962) Bijdrage tot het onderzoek naar de biosynthese van de tropa-alkaloiden. Pharm Weekbl 97:721–731

Srinivasan P, Smolke CD (2019) Engineering a microbial biosynthesis platform for de novo production of tropane alkaloids. Nat Commun 10(1):3634

Stewart C Jr et al (2013) Confluence of structural and chemical biology: plant polyketide synthases as biocatalysts for a bio-based future. Curr Opin Plant Biol 16(3):365–372

Suzuki K et al (1999) An Atropa belladonna hyoscyamine 6beta-hydroxylase gene is differentially expressed in the root pericycle and anthers. Plant Mol Biol 40(1):141–152

Tabor CW, Tabor H (1984) Polyamines. Annu Rev Biochem 53:749–790

Tavladoraki P et al (1998) Maize polyamine oxidase: primary structure from protein and cDNA sequencing. FEBS Lett 426(1):62–66

Tavladoraki P et al (2006) Heterologous expression and biochemical characterization of a polyamine oxidase from Arabidopsis involved in polyamine back conversion. Plant Physiol 141 (4):1519–1532

Tavladoraki P, Cona A, Angelini R (2016) Copper-containing amine oxidases and FAD-dependent polyamine oxidases are key players in plant tissue differentiation and organ development. Front Plant Sci 7:824

Teuber M et al (2007) Putrescine N-methyltransferases--a structure-function analysis. Plant Mol Biol 63(6):787–801

Torres M (2018) Structure and function of enzymes involved with the biosynthesis of benzylisoquinoline alkaloids. In: Graduate studies, science. University of Calgary, Calgary

Ullrich SF, Hagels H, Kayser O (2016) Scopolamine: a journey from the field to clinics. Phytochem Rev 16(2):333–353

Ushimaru R et al (2018) Substrate conformation correlates with the outcome of hyoscyamine 6beta-hydroxylase catalyzed oxidation reactions. J Am Chem Soc 140(24):7433–7436

Ushimaru R, Ruszczycky MW, Liu HW (2019) Changes in regioselectivity of H atom abstraction during the hydroxylation and cyclization reactions catalyzed by hyoscyamine 6beta-hydroxylase. J Am Chem Soc 141(2):1062–1066

Walton NJ, Robins RJ, Peerless AC (1990) Enzymes of N-methylputrescine biosynthesis in relation to hyoscyamine formation in transformed root cultures of Datura stramonium and Atropa belladonna. Planta 182(1):136–141

WHO (2019) World Health Organization model list of essential medicines: 21st list 2019. World Health Organization, Geneva

Wigle ID, Mestichelli LJJ, Spenser ID (1982) 2H n.m.r. spectroscopy as a probe of the stereochemistry of biosynthetic reactions. The biosynthesis of nicotine. J Chem Soc Chem Commun 12:662–664

Williams N et al (1977) Cocaine and metabolites - relationship between pharmacological activity and inhibitory action on dopamine uptake into striatal synaptosomes. Prog Neuropsychopharmacol 1(3-4):265–269

Wilson J et al (2018) The effects of the alkaloid scopolamine on the performance and behavior of two caterpillar species. Arthropod Plant Interact 12(1):21–29

Wink M (1998a) Modes of actions of alkaloids. In: Roberts MF, Wink M (eds) Alkaloids. Plenum Press, New York, pp 301–326

Wink M (1998b) A short history of alkaloids. In: Roberts MF, Wink M (eds) Alkaloids. Plenum Press, New York, pp 11–44

Xia K et al (2016) Promoting scopolamine biosynthesis in transgenic Atropa belladonna plants with pmt and h6h overexpression under field conditions. Plant Physiol Biochem 106:46–53

Yamada Y et al (1990) Biochemistry of alkaloid production in vitro. In: Charlwood BV, Rhodes MJC (eds) Secondary products from plant tissue culture. Oxford Science Publications, Oxford, pp 227–242

Yanagisawa H, Suzuki Y (1981) Corn agmatine iminohydrolase: purification and properties. Plant Physiol 67(4):697–700

Zhao K et al (2017) Enhancing tropane alkaloid production based on the functional identification of tropine-forming reductase in Scopolia lurida, a Tibetan medicinal plant. Front Plant Sci 8:1745

Zhao T et al (2019) Metabolic characterization of Hyoscyamus niger ornithine decarboxylase. Front Plant Sci 10:229

Ziegler J, Facchini PJ (2008) Alkaloid biosynthesis: metabolism and trafficking. Annu Rev Plant Biol 59:735–769

Plant Tropane Alkaloids: Commercial Stature and Production Developments

3

Shakti Mehrotra, Sonal Mishra, and Vikas Srivastava

Abstract

Prior to the advancements in synthetic chemistry and developments of new drug molecules, plants have been the richest and renewable source of medicinal compounds. A range of pharmaceutically important compounds are synthesized by plants as a result of their various metabolic activities. Tropane alkaloids (TAs) are well-known therapeutically active compounds produced by a number of plant families. The major bioactivity of TAs includes anticholinergic and stimulant properties due to which these compounds are frequently used in the drugs meant for the treatment of ailments like central nervous disorders, postoperative nausea, pain, motion sickness, stomach and intestinal disorders, Parkinson's disease, and cough. Besides, drugs used for localized and general anesthesia also contain tropane alkaloid as major ingredients. This huge commercial significance of plant TAs has created an ever-increasing commercial demand of these natural compounds. Although majority of demand is being met through traditional cultivation, the supply always remains unsatisfactory. Coupled with this, unsystematic harvesting and collection of plant resources to obtain these active molecules is creating an unpleasant scene from the environmental and ecological point of view. To combat this, several biotechnological strategies have also been adapted to produce plant TAs through in vitro cultures. The present chapter

S. Mehrotra
Department of Biotechnology, Institute of Engineering and Technology, Lucknow, Uttar Pradesh, India

S. Mishra
School of Biotechnology, University of Jammu, Jammu, Jammu and Kashmir, India

V. Srivastava (✉)
Department of Botany, Central University of Jammu, Samba, Jammu and Kashmir, India
e-mail: vikas.bot@cujammu.ac.in

V. Srivastava et al. (eds.), *Tropane Alkaloids*,
https://doi.org/10.1007/978-981-33-4535-5_3

provides a description on commercial stature and conventional/unconventional approaches for plant TA production.

Keywords

Tropane · In vitro cultures · Drugs · Pharmacology · Secondary metabolites · Solanaceae

3.1 Introduction

Plants are the rich repositories of a number of important bioactive phytochemical compounds. Normally, these compounds are secondary metabolites, production of which is triggered in plants during their exposure to various environmental stress factors. These phytochemicals have no direct or essential role in plant's life; however, they facilitate some physiological events like attracting pollinators and establishing mutual relationships with other organisms (Pagare et al. 2015). On the contrary, these phytochemicals have great significant value for humans and, thus, are considered as substantial material for a number of industrial purposes (Hussein and El-Anssary 2018). A large portion of these plant-based bioactive compounds is required in pharmaceutical industry where they are used as important ingredients either directly as crude or in composite form in various drug formulations (Thomford et al. 2018). About 25–30% of modern medicines for simple to serious ailments are derived from plant-based bioactive molecules, and this percentage is expected to increase up to 50% in times to come (Frank and Rene 2008). Besides, other industries like confectionery, cosmetics, textile, nutraceutical, and flavor and fragrance also depend on a significant contribution of plant-based secondary metabolites. Plant secondary metabolites are simple to complex structured molecules and are produced by plant cells via plagiaristic metabolic pathways apart from primary metabolic pathways. In current state of art, plant secondary metabolites have been categorized in alkaloids, terpenes, and phenolics as three main groups (Mehrotra et al. 2020). However, during the course of development and evolution of plant species, concurrent evolution and deviations in related biosynthetic pathways have led to the considerable structural modifications in these molecules. This has consequently resulted in the existence of a wide range of plant-derived bioactive molecules with enormous biological activities.

Alkaloids are considered as those biomolecules that contain at least one nitrogen atom in a heterocyclic ring. The name has been derived from their alkaline-like basic nature. Alkaloids that possess bicyclic amine with a pyrrolidine and a piperidine ring along with common nitrogen atom are commonly recognized as tropane alkaloids (Kohnen-Johannsen and Kayser 2019). Tropane alkaloids are highly bioactive and of immense pharmaceutical value due to which they are globally used as active ingredient of a large number of drug formulations.

The origin of using plants, plant parts, extracts, and their fractions goes back with the advent of human civilization. Plants have been used as a rich and renewable source of therapeutic or toxic formulations in various medicinal streams of ancient times. Due to therapeutic and toxic properties of these metabolites, the tropane alkaloid-bearing plants serve human kind for medicinal and other purposes (Lavecchia et al. 2013). The first documented reference to pharmacological and therapeutic properties of TA-producing plants of Solanaceae family dates back to 2000 BC from ancient Egypt and Mesopotamia. The document reveals the strong medicinal properties of *Hyoscyamus niger*, *Mandragora officinarum*, and *Atropa belladonna* and their use for medicinal purposes and deliberate poisoning of enemies and animals (de Pasquale 1984; Alizadeh et al. 2014). Medicinal properties of different Datura species are also well documented (Al-Snafi and Esmail 2017). These plants belong to the family Solanaceae in which several other members also possess narcotic, poisonous, and hallucinogenic properties due to the presence of specific tropane alkaloids. Therefore, in ancient times, these plants have a dominant use in medicine, rituals, magic, and superstitions associated with healing of various types of physical and mental disorders. Nowadays, attributing to these properties, tropane alkaloids and their native plants have been chemically explored for their pharmacological effects and, thus, utilized as a main ingredient in the various modern and traditional drugs meant for physical and mental ailments (Table 3.1).

3.2 Pharmacological Properties: Potential in the Pharmaceutical World

3.2.1 Anticholinergic Property

Basically, tropane alkaloids have anticholinergic properties attributing to the blocking activity of these compounds to the neurotransmitter molecule acetylcholine at synapses in the central and/or peripheral nervous systems (Kohnen-Johannsen and Kayser 2019). This property is in particular with three TAs that are atropine, hyoscyamine, and scopolamine which are native and very common to genus *Atropa* (deadly nightshade plants), *Hyoscyamus* (henbane plants), *Mandragora* (mandrake plants), *Latua* (sorcerer's tree), and *Datura* (jimson weed plant). Besides, the genus *Brugmansia* (angel trumpet plants) and plant members of the subfamily Solanoideae of Solanaceae such as the genus *Anisodus* are also known for producing all these acetylcholine-inhibiting TAs. Though the underground parts are recognized as site for synthesis of these metabolites, they also concentrate in leaves and seeds with varying degree of concentration. Upon oral intake of TAs in the human body, their direct absorption from the gut followed by rapid translocation into tissues to be metabolized through cytochrome P450-mediated oxidative methylation is reported. Normally, these TAs do not accumulate in the human body but are excreted as renal wastes (Adamse and van Egmond 2010; EFSA 2013).

As the possible causes for certain mental conditions like depression, anxiety and symptoms of epilepsy, Parkinson's, and Alzheimer's are associated with the dysfunction of muscarinic acetylcholine receptor system in the human body, these plant TAs with their antagonistic properties to muscarinic system are considered as

Table 3.1 Lead tropane alkaloids from plants and some of the common TA-based drugs used in modern medicine

Native plant genus	Tropane alkaloid	Bioactivity	Brand names	Drug utilization
Hyoscyamus *Scopolia* *Datura* *Anisodus* *Atropa* *Brugmansia* *Duboisia* *Mandragora*	Scopolamine	Anticholinergic Antispasmodic Mydriatic Cycloplegic	Isopto hyoscine Transderm-Scop Scopolamine Hydrobromide	Topical ophthalmic solution Transdermal patch Motion sickness and postoperative nausea
	Hyoscyamine	Anticholinergic Antispasmodic Antidote	Anaspaz Azuphen MB Hyoscyamus-30 (homeopathic formulation) Duboisinum mother tincture (homeopathic formulation) Soochi (ayurvedic formulation of *A. belladonna*)	Stomach and intestinal disorders, IBS, Parkinson's symptoms Discomfort, pain, spasms due to UTI, and other infections Treat mental and physical disorders Motion sickness, stomach disorders Pain killer in different sickness
	Anisodamine	Anticholinergic Antispasmodic	Anisodamine hydrochloride	Intestinal diseases
	Atropine	Anticholinergic Antispasmodic Antidote	AtroPen Atropine Diban cap Belladonna (homeopathic formulation) *Mandragora officinarum* (homeopathic formulation)	Nerve agent and insecticidal poisoning Preoperative treatment Sedative, bronchial spasm, asthma, whooping cough, cold, hay fever Treat constipation
Erythroxylum	Cocaine	Anesthetic Vasoconstrictor	Cocaine hydrochloride Goprelto	Topical anesthetic for oral, laryngeal, and nasal cavities Anesthetic for local application

potential drug compounds to these and other CNS-related serious ailments such as depression and anxiety (Jaffe 2013; Du et al. 2017). Scopolamine is able to pass through blood-brain barrier and is effective to create pupil dilation and reduce muscle spasms and motion sickness possibly by inhibiting neural functions from the inner ear to emetic center in the brain (Renner et al. 2005; Navarria et al. 2015). It also creates local anesthetic effect and is widely used as an ingredient of anesthetic drugs (Tsuchiya 2017). Hyoscyamine binds at muscarinic receptors in salivary, bronchial, sweat gland, eye, heart, and gastrointestinal tract which results in a reduction in salivary, bronchial, gastric, and sweat gland secretions; change in heart rate; and decreased gastrointestinal motility (Grynkiewicz and Grynkiewicz 2008). Hyoscyamine and related drugs have also been used to provide relief from heart problems, Parkinson's disease, respiratory illness, muscle and other spasms, and bowel illness including ulcers, colitis, irritable bowel syndrome, and pancreatitis (Grynkiewicz and Grynkiewicz 2008). Atropine is a stable compound and thus mostly preferred over hyoscyamine and others for being used in drug formulations (Kohnen-Johannsen and Kayser 2019). Both atropine and hyoscyamine are also used as antidotes against poisonous organophosphorous nerve agents like organophosphorous or carbamate insecticides. Anisodamine, belonging to the genus *Anisodus*, is used to improve and/or maintain blood flow during shock treatment. It is also used as an active ingredient in the drugs meant for blood circulatory and gastric disorders (Kohnen-Johannsen and Kayser 2019).

3.2.2 Stimulant Property

Some of the plant TAs such as cocaine and associated compounds are able to increase activity of the central nervous system and the body and, thus, used in the drugs that are pleasurable and invigorating or drugs that can mimic or modify the actions of endogenous catecholamines of the sympathetic nervous system (sympathomimetic effects). Major tropane alkaloid in this category is cocaine naturally obtained from the genus *Erythroxylum* (coca plants) where it is synthesized in green parts of the plant unlike other TAs which are normally synthesized in roots and then transported to aerial parts (Restrepo et al. 2019). In a study, identification of cocaine-synthesizing enzyme activity in leaf extracts of coca, purification and isolation of related protein, and also the cloning of corresponding gene revealed existence of an entirely different enzyme aldo-keto reductase (AKR) in coca plants in comparison with other Solanaceae members (Huo et al. 2018). The activities of AKR were found high in aerial parts especially in leaves in comparison to roots of coca plants. The results in this study were not in agreement with results obtained from other TA-producing plants where maximum activities of TA pathway enzymes occur in root tissues. This has led to the conclusion that TA biosynthesis in Erythroxylaceae and Solanaceae are not associated with each other and evolved independently (Jirschitzka et al. 2012). Cocaine, the major alkaloid of this family, is considered as drug of use and abuse both because on one hand it is clinically used as local anesthetic and vasoconstrictor while on the other hand, it is a powerful nervous stimulant and highly addictive as it strongly influences dopaminergic pathway in the brain. Owing to these properties, it is frequently used to increase attentiveness, talkativeness, competence, energy and motor activity, feeling of euphoria, happiness, and sexuality (Nestler 2005; Porrino et al. 2007). Anesthesia caused by cocaine is a result of blocking conduction in peripheral nerves by inactivating sodium channels. Sodium influx through these channels is necessary for the depolarization of nerve cell membranes and subsequent propagation of impulses along the course of the nerve. Besides cocaine, ecgonine (precursor of cocaine), benzoylecgonine (metabolite of cocaine), and methylecgonine (another natural coca TA) are the compounds that have similar bioactivities. Another TA, i.e., hydroxytropacocaine, from coca plant is identified and isolated that has reported zero activity (Restrepo et al. 2019). Owing to such high medicinal value of these compounds, several drugs and drug formulations are being prepared using these compounds as main ingredient. These drugs are used for several ailments in different medicinal systems including Ayurveda, allopathy, and homeopathy (Table 3.1). A continuous commercial demand of plant TAs in different medicinal systems has given a center stage position to these compounds and an ever-increasing market value. In a report, provided by Coherent Market Insights Analysis (2018), the key pharmaceutical companies and organizations operating in global tropane alkaloid market with their prominent TA-based drugs and other formulations include GlaxoSmithKline plc., Novartis AG, Alchem International, Alkaloids of Australia, Fine Chemicals Corporation, and Baxter International Inc. The report also suggests that in 2017 global scopolamine market size is expected to witness a compound annual growth rate (CAGR) of 4.8% over the forecast period (2018–2026) which was valued at US$ 370.4 million in 2017. Similarly, in case of atropine, the global market will observe a CAGR of 1.7% during 2021–2026. At present, global atropine market is valued at 530.2 million US$ which is expected to touch 469.6 million US$ by 2026 (https://www.marketwatch.com/press-release/atropine-market-2020-global-industry-trends-share-size-demand-growth-opportunities-industry-revenue-future-and-business-analysis-by-forecast-2026-2020-06-04).

Such a huge market value of plant TAs has led to their commercial-scale production worldwide. Although the traditional cultivation solely cannot meet the commercial demand, therefore, synthetic derivatives of these alkaloids came into existence. Additionally, the advent of concepts of structure-activity relationship (SAR) and quantitative SAR (QSAR), pharmacophore, ligand-receptor binding, and other aspects of computational designing has facilitated the chemical synthesis of lead derivative compounds based on natural TAs. This has also thriven the effective drug development through the exploration of structure-based enhanced bioactivity of new chemical analogs of prototype compounds. In this regard, tropeines, tropisetron, benztropine, and homatropine are the few names of synthetic TA analog compounds of clinical applications. A detailed account on TAs and their synthetic derivatives as drug compounds has been given in article presented by Grynkiewicz and Gadzikowska (2008).

Despite the development of synthetic production, still at present, majority of plant TAs utilized by pharmaceutical companies is produced agriculturally through field cultivation of TA-producing species (Table 3.2). Various species of the genus *Duboisia*, *Atropa*, *Scopolia*, and *Hyoscyamus* are cultivated worldwide for large-scale production of scopolamine, hyoscyamine, and other related TAs. Species of *Brugmansia* have been cultivated in Europe since mid-1990s and known to produce 0.8% scopolamine annually from dried leaves (Levy 1977). The Australian species *B. candida* hybrid is also known as a potential source of scopolamine with highest yields of 0.56% from young dried leaves (Griffin and Lin 2000). Two species of *Duboisia*, i.e., *D. myoporoides* and *D. leichhardtii*, are known to be cultivated in their natural locations of Southwest Queensland. In recent years, *D. myoporoides* is grown little and is not a preferred alkaloid source because of its pyridine base content

Table 3.2 Geographical distribution of commercially grown TA-producing plant species and their total alkaloid yields

Plant	Geography	Alkaloid yields
A. belladonna *A. baetica*	Europe, North America, and Western Asia	In a range of 0.34–0.68% DW
H. muticus *H. niger* *H. albus*	North Africa, Western Asia, Europe, India	0.04–0 0.72% DW
Datura metel *D. innoxia* *D. leichhardtii* *D. stramonium*	America, Southeast Asia, Australia	1.2–1.8% DW
Duboisia myoporoides *D. Leichhardtii hybrid*	Australian rainforests, Japan	High up to 7%
Scopolia japonica	Europe and Asia	
Erythroxylum coca	South America particularly in northern Brazil, Bolivia, and Peru	0.3–1.5%, averaging 0.8%

which makes scopolamine isolation little less efficient. Rather, a hybrid of *D. myoporoides* and *D leichhardtii* is being cultivated. The leaves of this commercially explored hybrid plant contains scopolamine (1.54%) and hyoscyamine (0.1%) (Griffin and Lin 2000). The alkaloid content of cultivated crops depends upon a number of factors including age of plant, temperature, humidity, sunlight intensity and shade duration, and seasonal variations (Ikenaga et al. 1985). Also, exposure to biotic factors such as weeds and insects also influences the productivity of plants. Not only the climate but also the soil composition and added fertilization have a significant impact on the alkaloid biosynthesis as well as in plant development. *Datura* plants supplemented with fertilizers exhibited higher hyoscyamine and scopolamine level. Likewise, different species of *Hyoscyamus* demonstrated a significant correlation between N, P, and K supplementation and soil, and their hyoscyamine and scopolamine contents however are contrary to this; *Duboisia* plants during field cultivation to produce scopolamine did not show any correlation in alkaloid content and fertilizer supplementation (Al-Humaid 2004; Alaghemand et al. 2013). In a nutshell, agricultural production of TAs depends upon optimization of several environmental, physical, and chemical factors which at majority of times do not synchronize to give optimum yields. There are still a lot of efforts required to improve large-scale production of plant TAs through traditional agriculture method. In a very first step, selection and screening of high-yielding clones and their germplasm conservation are required. Then there is a need of systemic trials for growing TA-producing elite plants under strictly controlled and reproducible conditions. This will help to understand the effect of various environmental factors on plant biomass and alkaloid yield. The knowledge obtained here could be utilized to field-grown plants to achieve optimal productivity.

3.3 Sustainable TA Production: Biotechnological Intervention

In spite of worldwide agricultural production of plant TA, the commercial demand remains unmet. Further, a long maturation time of plants to produce harvestable biomass coupled with unsystematic cultivation and climatic and environmental limitations has opened the direction of unconventional production of plant TAs through biotechnological interventions. Biotechnology has facilitated TA production in various ways. Not only this, it has also provided a flawless method to propagate, maintain, and conserve the germplasm of elite and endangered clones of TA-producing plants. Various details of biotechnological interventions for plant TA production has been given in upcoming chapters of this book; however, an introduction to various biotechnological strategies utilized for TA production is being presented in ensuing text and Fig. 3.1.

Fig. 3.1 Various biotechnological strategies explored in TA-producing plants for multiplication and metabolite production

3.4 In Vitro Multiplication, Germplasm Conservation, and Generation of Biochemical Variants of TA-Producing Plants

In vitro multiplication is a method that not only facilitates conservation of elite and endangered species but also creates an in vitro repository of plants of commercial importance and introduces significant variations in morphological and biochemical significance through somaclonal variations. A number of TA-producing plant species from different genus have been subjected to in vitro multiplication for their conservation and other biotechnological purposes. An in vitro propagation method has been developed for multiplication and conservation of *Atropa acuminata* Royle ex Lindl—an Indigenous, threatened medicinal plant—through shoot tips and nodal explants (Ahuja et al. 2002). Plants of *A. belladonna* that were regenerated from callus cultures also reveal enhanced morphological and biochemical variability. The most valuable individual clones were in vitro propagated and used in the advanced program to obtain new cultivars with high TA-yielding traits (Toth et al. 1991). In a comparative study involving *Hyoscyamus aureus* wild and in vitro-grown plants, significant variations in tropane alkaloids contents were observed. This method offers a TA production method from comparatively less studied species of the genus *Hyoscyamus* (Besher et al. 2014). Considering the unsystematic exploitation of important medicinal plants as cause of depleting their gene pool, in a recent study, genetic variability through targeted mutations was introduced in *H. niger* (Shah et al. 2020). The study targeted putrescine N-methyltransferase (PMT) and 6β-hydroxy hyoscyamine (H6H) genes of alkaloid biosynthetic pathway and development of a reproducible protocol for in vitro multiplication and mutagenesis. The study was effective in developing variations not only at phenotypic level but also in overexpression of targeted genes and enhanced secondary metabolite content of this important plant. Further, the study also proposes this method of inducing genetic variability through targeted mutations as potential method for generating varieties with increased content of active compounds of other related or unrelated TA-bearing plants (Liu et al 2010). Polyploidy is a general phenomenon in nature and known to be the cause of high degree of morphological, physiological, and biochemical variations. Keeping this in mind and considering the significantly changed alkaloid profile of diploid and autotetraploid plants of D. *stramonium*, in another study, effect of ploidy level and culture conditions on in vitro-grown *H. muticus* was also studied on the production of tropane alkaloids (Berkov and Philipov 2002; Dehgan et al. 2012). The tetraploidy significantly affected the growth rate and generated variations in alkaloid accumulation patterns as tetraploid plants, despite lower biomass production, produced approximately 200% greater scopolamine than that of diploid plants under similar growth conditions (Dehgan et al. 2012). Although the hyoscyamine is the main alkaloid in the *H. muticus* plants, manipulation of ploidy level and culture conditions successfully changed the scopolamine/hyoscyamine ratio toward scopolamine. Conversion of hyoscyamine into scopolamine is very important due to the higher market value of scopolamine.

3.5 Plant Cell Suspension and Hairy Root Cultures

In a classical study, *A. belladonna* callus cultures were established from different explant sources in which atropine was detected in root-borne callus (West and Mika 1957). From now on, several approaches have been employed to utilize callus and/or suspension cultures of TA-producing plants for the production of these pharmaceutically important metabolites. The question of TA synthesis site was a subject of confusion till the discovery of the root pericycle-specific location of the PMT and H6H (Hashimoto et al. 1991; Suzuki et al. 1999). This may be the main reason of aiming roots and root-borne cultures as main source of alkaloids in these plants. Establishment of callus and suspension cultures for TA production is advantageous in terms of potential upscaling and ease of extraction; however, on the other hand, in a scale-up bioprocess, the overall production cost and other hitches associated with optimization of process parameters generally make this aspect practically complicated. Another limitation associated with callus and suspension cultures is the cell-specific compartmentalization during biosynthesis of major TAs. The underground parts of the plants are the major site for TA biosynthesis, and this cell-specific compartmentalization in TA during biosynthesis is evidenced through the studies that confirmed expression of major TA biosynthetic enzymes in different root tissue. The majority of enzymes like TR-I and TR-II are known to be expressed in root endodermal and cortex cells, respectively. Expression of PMT and H6H is known to be normally present in root pericycle cells. Thus, an undifferentiated mass of cells (callus) may have limitations to produce desired amount of TAs. Additionally, the genetic instability of these cells is also considered as limiting factor during their use as production platform for TAs. Maybe these are the few reasons that till now no callus and/or suspension culture of any TA-bearing plant is successfully utilized for their large-scale production.

Talking about hairy root cultures (Srivastava et al. 2018, 2020) for TA production is a matter of extensive exploration, and without which success story of biotechnological intervention of TA production remains incomplete (Srivastava et al. 2017a; Skała and Sitarek 2020). A complete chapter (Chap. 6) of this book is committed to discuss the hairy root cultures and their contribution to TA production. Though still restricted to bench scale, the large success of hairy root cultures for TA production is indebted to few reasons. These reasons are briefly pointed out below with relevant and/or recent references. A survey of cross-references of the literature cited here will provide classical examples of each study related to hairy root-based TA production.

1. Ease of establishing *Agrobacterium*-based genetic transformation protocol because Solanaceae (dominating TA-producing plant family) plant members are rich in plant-derived phenolic compounds that are inducers of *vir* genes in bacterium such as acetosyringone upon wounding, be it natural or deliberate under laboratory conditions (Subramoni et al. 2014; Mehrotra et al. 2015; Gebhardt 2016; Hwang et al. 2017).
2. High susceptibility to a range of bacterial strains (Zehra et al. 1999; Akramian et al. 2008; Shakeran et al. 2014).

3. Fast-growing high biomass-yielding hairy roots do not require specific growth hormone or any other culture condition (Mehrotra et al. 2015; Srivastava et al. 2017a, b).
4. Significantly responsible (in terms of increased production of metabolites) to macro- and micro-nutrient alteration/supplementation in the culture media (Pinol et al. 1999; Pudersell et al. 2007, 2012; Harfi et al. 2016).
5. Genetic and biochemical stability during long-term culture (Maldonado-Mendoza et al. 1993; Marconi et al. 2008).
6. Responsive to a range of biotic and abiotic elicitors (Goel et al. 2011; Moharrami et al. 2017; Shakeran et al. 2017; Khezerluo et al. 2018).
7. Suitable platform for TA yield enhancement through single- and/or multistep gene modification methods (Mehrotra et al. 2010; Kai et al. 2011, 2012; Zhao et al. 2017).
8. Ease of upscaling (Cardillo et al. 2010; Jaremicz et al. 2014; Habibi et al. 2015; Mehrotra and Srivastava 2017).
9. Apart from this, TA-producing HRCs have also been utilized for certain non-native metabolite-based biotechnological applications (Srivastava et al. 2012, 2013, 2017b; Srivastava 2015; Mehrotra and Srivastava 2017).

The establishment of a high TA-yielding protocol through growing hairy roots in a bioreactor for industrial-scale production is a challenging task and in need of immense scientific and engineering attention (Mehrotra et al 2016). Alternatives to the agricultural production of TAs are being persistently continued. Be it chemical synthesis or through biotechnological production, the limitation surrounds over complex interactions and numerous factors influencing the tropane alkaloid biosynthesis. Therefore, the major attention should be given to the optimization of process in terms of standardized parameters for desirable production. Further, there is a need of stable and reproducible production system, which also needs to bring along the necessary characteristics for a scale-up to industrial levels for commercialization. Until now, these requirements are not fully met, and further studies are ongoing.

References

Adamse P, van Egmond HP (2010) Tropane alkaloids in food. RIKILT—Institute of Food Safety, Wageningen. https://www.researchgate.net/publication/254834358

Ahuja A, Sambyal M, Koul S (2002) In vitro propagation and conservation of Atropaacuminata royle ex Lindl—an indigenous threatened medicinal plant. J Plant Biochem Biotechnol 11:121–124. https://doi.org/10.1007/BF03263148

Akramian M, Tabatabaei SMF, Mirmasoumi M (2008) Virulence of different strains of agrobacterium rhizogenes on genetic transformation of four Hyoscyamus species. Amer-Euras J Agric Environ Sci 3:759–763

Alaghemand A, Ghorbanpour M, Asli DE et al (2013) Influence of urea fertilization on tropane alkaloids content of Henbane (Hyoscyamus niger L.) under hydroponic cultureconditions. Adv Environ Biol 7(2):301–307

Al-Humaid AI (2004) Effects of compound fertilization on growth and alkaloids of Datura plants. J Plant Nutr 27(12):2203–2219

Alizadeh A, Moshiri M, Alizadeh J, Balali-Mood M (2014) Black henbane and its toxicity–a descriptive review. Avicenna J Phytomed 4(5):297

Al-Snafi, Esmail A (2017) Medical importance of Datura fastuosa (syn: Datura metel) and Datura stramonium—a review. IOSR J Pharm 7(2):43–58. https://doi.org/10.9790/3013-0702014358

Berkov S, Philipov S (2002) Alkaloid production in diploid and autotetraploid plants of Datura stramonium. Pharm Biol 40(8):617–621

Besher S, Al-Ammouri Y, Murshed R (2014) Production of tropan alkaloids in the in vitro and callus cultures of Hyoscyamus aureus and their genetic stability assessment using ISSR markers. Physiol Mol Biol Plants 20(3):343–349. https://doi.org/10.1007/s12298-014-0242-6

Cardillo AB, Otálvaro AAM, Busto VD et al (2010) Scopolamine, anisodamine and hyoscyamine production by Brugmansia candida hairy root cultures in bioreactors. Process Biochem 45 (9):1577–1581

Coherent Market Insights Analysis (2018) https://www.coherentmarketinsights.com/market-insight/scopolamine-market-1965

de Pasquale A (1984) Pharmacognosy: the oldest modern science. J Ethnopharmacol 11(1):1–16

Dehghan EH, Oksman-Caldentey S, Ahmadi K-M, Ahmadi FS (2012) Production of tropane alkaloids in diploid and tetraploid plants and in vitro hairy root cultures of Egyptian henbane (Hyoscyamus muticus L.). Plant Cell Tiss Org Cult 110:35–44. https://doi.org/10.1007/s11240-012-0127-8

Du N, Liu Y, Zhang X et al (2017) Discovery of new muscarinic acetylcholine receptor antagonists from Scopoliatangutica. Sci Rep 7:46067. https://doi.org/10.1038/srep46067

EFSA (2013) Scientific opinion on tropane alkaloids in food and feed. EFSA J 9:1–134

Frank P, Rene A (2008) Natural compounds as drugs, vol 1. Bikhauser, Boston, p 350

Gebhardt C (2016) The historical role of species from the Solanaceae plant family in genetic research. Theor Appl Genet 129(12):2281–2294. https://doi.org/10.1007/s00122-016-2804-1

Goel MK, Mehrotra S, Kukreja AK (2011) Elicitor-induced cellular and molecular events areresponsible for productivity enhancement in hairy root cultures: an insight study. Appl Biochem Biotechnol 165:1342–1355

Griffin WJ, Lin GD (2000) Chemotaxonomy and geographical distribution of tropane alkaloids. Phytochemistry 53:623–637

Grynkiewicz G, Gadzikowska M (2008) Tropane alkaloids as medicinally useful natural products and their synthetic derivatives as new drugs. Pharmacol Rep 60(4):439–463

Habibi P, Piri K, Ali D, Moghadam YA, Ghiasvand T (2015) Increasing Scopolamine content in hairy roots of Atropa belladonna using bioreactor. Braz Arch Biol Technol 58:2

Harfi B et al (2016) Hyoscyamine production in hairy roots of three Datura species exposed to high-salt medium. In Vitro Cell Dev Biol 52:92–98

Hashimoto T, Hayashi A, Amano Y et al (1991) Hyoscyamine 6b-hydroxylase, an enzyme involved in tropane alkaloidbiosynthesis, is lokalized at the pericycle of the root. J Biol Chem 266 (7):4648–4653

Huo J, Du B, Sun S, He S, Zhao N, Liu Q, Zhai H (2018) A novel aldo-ketoreductase gene. Front Agri Sci Eng 5(2):206–213

Hussein RA, El-Anssary AA (2018) Plants secondary metabolites: the key drivers of the pharmacological actions of medicinal plants. In: Builders PH (ed) Herbal medicine. Intech Ope, London

Hwang HH, Yu M, Lai EM (2017) Agrobacterium-mediated plant transformation: biology and applications. Arabidopsis 15:e0186. https://doi.org/10.1199/tab.0186

Ikenaga T, Hama K, Takemoto S et al (1985) Seasonal variations of tropane alkaloids content in Duboisia. Jpn J Trop Agr 29(4):229–230

Jaffe RJ, Novakovic V, Peselow ED (2013) Scopolamine as an Antidepressant. Clin Neuropharmacol 36(1):24–26

Jaremicz Z, Luczkiewicz M, Kokotkiewicz A et al (2014) Production of tropane alkaloids in Hyoscyamus niger (black henbane) hairy roots grown in bubble-column and spray bioreactors. Biotechnol Lett 36(4):843–853

Jirschitzka J, Schmidt GW, Reichelt M, Schneider B, Gershenzon J, D'Auria JC (2012) Plant tropane alkaloid biosynthesis evolved independently in the Solanaceae and Erythroxylaceae. Proc Natl Acad Sci U S A 109(26):10304–10309

Kai G, Yang S, Luo X et al (2011) Co-expression of AaPMT and AaTRI effectively enhances the yields of tropane alkaloids in Anisodusacutangulus hairy roots. BMC Biotechnol 11:43

Kai G, Zhang A, Guo Y, Li L, Cui L, Luo X, Xiao J (2012) Enhancing the production of tropane alkaloids in transgenic Anisodusacutangulus hairy root cultures by over-expressing tropinonereductase I and hyoscyamine-6β-hydroxylase. Mol Bio Syst 8(11):2883–2890

Khezerluo M, Hosseini B, Amiri J (2018) Sodium nitroprusside stimulated production of tropane alkaloids and antioxidant enzymes activity in hairy root culture of Hyoscyamus reticulatus L. Acta Biol Hung 69(4):437–448

Kohnen-Johannsen KL, Kayser O (2019) Tropane alkaloids: chemistry, pharmacology, biosynthesis and production. Molecules 24(4):796

Lavecchia T, Giuseppina R, Amina A, Maria TG (2013) Healthy and adverse effects of plant-derived functional metabolites: the need of revealing their content and bioactivity in a complex food matrix. Crit Rev Food Sci Nutr 53:198–213

Levy R, (1977) Jimson seed poisoning—A new hallucinogen on the horizon. J Am Coll Emerg Physicians 6(2):58–61

Liu X, Yang C, Chen M, Li M, Liao Z, Tang K (2010) Promoting scopolamine accumulation in transgenic plants of *Atropa belladonna* generated from hairy roots with over expression of *pmt* and *h6h* gene. J Med Plant Res 4(17):1708–1713

Maldonado-Mendoza IE, Ayora-Talavera T, Loyola-Vargas VM (1993) Establishment of hairy root cultures of Datura stramonium. Characterization and stability of tropane alkaloid production during long periods of subculturing. Plant Cell Tiss Org Cult 33:321–329. https://doi.org/10.1007/BF02319018

Marconi P, Setten L, Cálcena E, Alvarez M, Pitta S (2008) Changes in growth and tropane alkaloid production in long-term culture of hairy roots of Brugmansia candida. J Integr Biosci 3:38

Mehrotra S, Srivastava V (2017) Hairy root in vitro systems: a suitable biological matrix for plant based remediation of environmental pollutants. In: Anjum NA (ed) Chemical pollution control with microorganisms. Nova Science, New York, pp 247–276

Mehrotra S, Rahman LU, Kukreja AK (2010) An extensive case study of hairy-root cultures forenhanced secondary-metabolite production through metabolic-pathway engineering. Biotechnol Appl Biochem 56:161–172

Mehrotra S, Srivastava V, Ur Rahman L, Kukreja AK (2015) Hairy root biotechnology-indicativetimeline to understand missing links and future outlook. Protoplasma 252 (5):1189–1201

Mehrotra S, Mishra S, Srivastava V (2016) Bioreactor technology for hairy roots cultivation. In: Pavlov A, Bley T (eds) Bioprocessing of plant in vitro systems, reference series in phytochemistry. Springer, Berlin/Heidelberg, pp 91–114

Mehrotra S, Mishra S, Srivastava V (2020) Hairy roots biotechnology unzipped: a journey of reality and promises. In: Srivastava V et al (eds) Hairy root cultures based applications, rhizosphere biology. Springer, Singapore

Moharrami F, Hosseini B, Sharafi A, Farjaminezhad M (2017) Enhanced production of hyoscyamine and scopolamine from genetically transformed root culture of Hyoscyamus reticulatus L. elicited by iron oxide nanoparticles. In Vitro Cell Dev Biol Plant 53(2):104–111

Navarria A et al (2015) Rapid antidepressant actions of scopolamine: role of medial prefrontal cortex and M1-subtype muscarinic acetylcholine receptors. Neurobiol Dis 82:254–261

Nestler EJ (2005) The neurobiology of cocaine addiction. Sci Pract Perspect 3(1):4–10. https://doi.org/10.1151/spp05314

Pagare S, Bhatia M, Tripathi N, Pagare S, Bansal YK (2015) Secondary metabolites of plants and their role: overview. Curr Trends Biotechnol Pharm 9:293–304

Pinol MT, Palazon J, Cusido RM, Ribo M (1999) Influence of calcium ion concentration in the medium on tropane alkaloid accumulation in Datura stramonium hairy roots. Plant Sci 141:41–49

Porrino LJ, Smith HR, Nader MA, Beveridge TJ (2007) The effects of cocaine: a shifting target over the course of addiction. Prog Neuropsychopharmacol Biol Psychiatry 31(8):1593–1600. https://doi.org/10.1016/j.pnpbp.2007.08.040

Pudersell K, Vardja R, Vardja T, Arak E, Raal A (2007) Influence of calcium on the tropane alkaloid production and riboflavine excretion in the root cultures of Hyoscyamus niger L. Eur J Pharm Sci 32(Suppl):S33–S34

Pudersell K, Vardja T, Vardja R, Matto V, Arak E, Raal A (2012) Inorganic ions in the medium modify tropane alkaloids and riboflavin output in Hyoscyamus niger root cultures. Pharmacogn Mag 8(29):73–77. https://doi.org/10.4103/0973-1296.93330

Renner UD, Oertel R, Kirch W (2005) Pharmacokinetics and pharmacodynamics in clinical use of scopolamine. Ther Drug Monit 27:655–665

Restrepo DA, Saenz E, Jara-Muñoz OA, Calixto IF, Rodríguez-Suárez S, Zuleta P, Chavez BG, Sanchez JA, D'Auria JC (2019) Erythroxylum in focus: an interdisciplinary review of an overlooked genus. Molecules 24:3788. https://doi.org/10.3390/molecules24203788

Shah D, Kamili AN, Wani AA, Majeed U, Wani ZA, Sajjad N et al (2020) Promoting the accumulation of scopolamine and hyoscyamine in *Hyoscyamus niger* L. through EMS based mutagenesis. PLoS One 15(5):e0231355

Shakeran Z, Keyhanfar M, Gholamreza A (2014) Hairy roots formation in four Solanaceae species by different strains of agrobacterium rhizogenes. J Med Plants By-prod 2:155–160

Shakeran Z, Keyhanfar M, Ghanadian M (2017) Biotic elicitation for scopolamine production by hairy root cultures of Datura metel. Mol Biol Res Commun 6(4):169–179

Skała E, Sitarek P (2020) Hairy roots as a source of tropane alkaloids. In: Srivastava V, Mehrotra S, Mishra S (eds) Hairy root cultures based applications. Rhizosphere biology. Springer, Singapore. https://doi.org/10.1007/978-981-15-4055-4_2

Srivastava V (2015) Studies on hairy roots as bioreactors for normal and elicited biotransformationof exogenous molecules. Ph.D thesis, University of Lucknow, Lucknow

Srivastava V, Negi AS, Ajayakumar PV, Khan SA, Banerjee S (2012) Atropa belladonna hairyroots: orchestration of concurrent oxidation and reduction reactions for biotransformation ofcarbonyl compounds. Appl Biochem Biotechnol 166(6):1401–1408

Srivastava V, Kaur R, Chattopadhyay SK, Banerjee S (2013) Production of industrially importantcosmaceutical and pharmaceutical derivatives of betuligenol by Atropa belladonna hairy rootmediated biotransformation. Ind Crop Prod 44:171–175

Srivastava V, Mehrotra S, Verma PK (2017a) Biotechnological interventions for production oftherapeutic secondary metabolites using hairy root cultures of medicinal plants. In: Current developments in biotechnology and bioengineering; crop modification, nutrition, and food production. Elsevier, Amsterdam, pp 259–282

Srivastava V, Mehrotra S, Mishra S (2017b) Biotransformation through hairy roots: perspectives, outcomes, and major challenges. In: Jha S (ed) Transgenesis and secondary metabolism. Springer, Cham, pp 1–24

Srivastava V, Mehrotra S, Mishra S (2018) Hairy roots: an effective tool of plant biotechnology. Springer, Cham

Srivastava V, Mehrotra S, Mishra S (2020) Hairy root cultures based applications. Springer, Cham

Subramoni S, Nathoo N, Klimov E, Yuan ZC (2014) Agrobacterium tumefaciens responses to plant-derived signaling molecules. Front Plant Sci 5:322. https://doi.org/10.3389/fpls.2014.00322

Suzuki K, Yamada Y, Hashimoto T (1999) Expression of Atropabelladonna putrescine N-methyltransferase gene in rootpericycle. Plant Cell Physiol 40(3):289–297

Thomford NE, Senthebane DA, Rowe A, Munro D, Seele P, Maroyi A, Dzobo K (2018) Natural products for drug discovery in the 21st century: innovations for novel drug discovery. Int J Mol Sci 19:1578

Tóth ET, Onisei T, Amariei D, Lazurca D (1991) Variability in tissue culture regenerated plants of Atropa belladonna. Acta Hortic 289:269–272

Tsuchiya H (2017) Anesthetic agents of plant origin: a review of phytochemicals with Anesthetic activity. Molecules 22:1369. https://doi.org/10.3390/molecules22081369

West FR Jr, Mika ES (1957) Synthesis of atropine by isolated roots and root-callus cultures of belladonna. Bot Gaz 119(1):50–54

Zehra M, Banerjee S, Sharma S, Kumar S (1999) Influence of agrobacterium rhizogenes strains on biomass and alkaloid productivity in hairy root lines of Hyoscyamus muticus and H. albus. Planta Med 65(1):60–63

Zhao K, Zeng J, Zhao T et al (2017) Enhancing tropane alkaloid production based on the functional identification of Tropine-forming reductase in Scopolialurida, a Tibetan medicinal plant. Front Plant Sci 8:1745

Part II

Tropane Alkaloids: In Vitro and Allied Interventions

Involvement of Various Biotechnological Contrivances for Tropane Alkaloid Biosynthesis and Applications of Tropane Alkaloid-Bearing In Vitro Cultures

4

Vikas Srivastava, Sonal Mishra, Skalzang Lhamo, Aksar Ali Chowdhary, and Shakti Mehrotra

Abstract

Plant natural metabolites (particularly secondary metabolites, SMs) promise solutions for many human ailments, since ancient time. With the advancement of plant biochemistry and phytochemistry along with its association with bioprospection studies, many metabolites have become qualified as specific (natural or their derivatives) molecule for targeted therapies. SM comprises a broad array of metabolites (viz., phenolics, terpenes, alkaloids etc.) and has specific biosynthesis and a very unique contribution in plants' life, though it is considered non-vital. Among SMs, alkaloids are the class of nitrogenous organic compounds that have further classification into TA (tropane alkaloid), TIA (terpene indole alkaloid), BIA (benzyl isoquinoline alkaloid), etc. Tropane alkaloids (TAs) are a class of bicyclic alkaloids, which contain a tropane ring in their chemical structure and represent in some of the angiospermic families, especially Solanaceae and Erythroxylaceae. Considering their demand and conservation issues about the management of natural resources harboring them, the interventions of sustainable and long-term production solutions are highly advocated. In order to highlight these issues, the present chapter will address the in vitro options as alternative production system, enhancement strategies, and allied applications of tropane alkaloid-bearing in vitro cultures.

Vikas Srivastava and Sonal Mishra contributed equally to this work.

V. Srivastava · S. Lhamo · A. A. Chowdhary
Department of Botany, Central University of Jammu, Samba, Jammu and Kashmir, India

S. Mishra
School of Biotechnology, University of Jammu, Jammu, Jammu and Kashmir, India

S. Mehrotra (✉)
Department of Biotechnology, Institute of Engineering and Technology, Lucknow, Uttar Pradesh, India

V. Srivastava et al. (eds.), *Tropane Alkaloids*,
https://doi.org/10.1007/978-981-33-4535-5_4

Keywords

Secondary metabolites · Elicitation · Metabolic engineering · Bioreactor · Molecular farming

4.1 Introduction

With the realization of idea that the biological systems or parts thereof can be directly prospected for valued products for environmental and human welfare (biotechnology), unabated efforts are being made in this direction (Canter et al. 2005; Hunter 2011). The unsystematic harvest of plants as natural resources for native metabolites causes perturbations at environmental and ecological level, and a genuine concern about this has created a foundation to explore the alternative methods for production, simultaneously maintaining the natural homeostasis. In the most talked about alternative method of plant cell and tissue culture, the initial experimentations generously advocated and proved that plants can be cultured under artificial and controlled conditions when supplemented with basal growth medium and phytohormones (Oseni 2018). The technique has attained recognition of full-fledged technology for plant conservation and multiplication in previous century. Simultaneously, a combination of this technology of growing plants in vitro with phyto- and biochemistry had developed a promising research platform of "plant in vitro culture-mediated secondary metabolite production" (Espinosa-Leal et al. 2018; Chandran et al. 2020). Since then, much advancement to this technology has led to the development of novel culture (production) systems such as callus, cell suspension, shoot cultures, and adventitious and hairy root cultures. With time, various scientific studies aimed to optimize these production systems for desired scale metabolite production in order to establish a commercial-scale production platform. Out of various types of "in vitro cultures" explored today, the cell suspension and hairy root cultures have emerged as most dominating production systems used enormously for SM production (Srivastava et al. 2018; Gutierrez-Valdes et al. 2020). These two systems are also reported to be frequently used worldwide for producing plant tropane alkaloids (TAs; Skala and Sitarek 2020). Plant tropane alkaloids are secondary metabolites of high medicinal importance. Range of tropane alkaloids including hyoscyamine, scopolamine, atropine, anisodamine, etc. are produced by plants of different angiosperm families (Fig. 4.1). A brief text is given below that highlights few of the instances as examples for the success story of plant tissue culture as alternative production system for TA production. Additionally, the text will also throw light on some of the wonderful allied biotechnological applications offered by "in vitro cultures" of important tropane alkaloid-bearing plants.

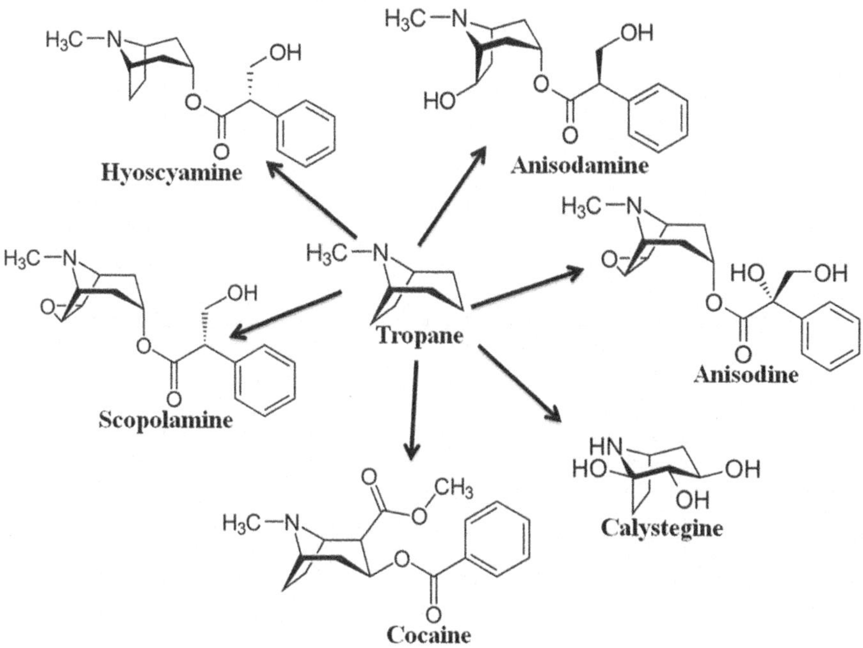

Fig. 4.1 Structural details of major tropane alkaloids

4.2 In Vitro Cultures as Alternative TA Production System

The TA production has been achieved mainly in callus and cell suspension cultures and hairy root cultures (HRCs, a cultured syndrome after deliberate *A. rhizogenes* infection; Mehrotra et al. 2015; Table 4.1). Among them considering the maintenance cost, the former two also require phytohormones in some or all phases of induction and/or maintenance; however, in contrast, HRCs (Srivastava et al. 2018, 2020) do not need any supplementation of phytohormones and/or growth regulators. Therefore, HRCs are considered as feasible option from economic point of view, and superimpose other in vitro TA production systems in recent decades. However, due to ease of upscaling in larger culture vessels, the suspension culture has also been widely explored for TA production. At present, hyoscyamine, scopolamine, atropine, and cocaine are some of the major tropane alkaloids with high therapeutic value that are produced through suspension and HRCs of TA-bearing plants.

Plenty of literature is available on initiation and exploration of callus and cell suspension culture of several TA-bearing plants (Yamada and Hashimoto 1982; Yamada and Endo 1984; Kinsara and Seif El-Naser 1990; Herouart et al. 1991; Gontier et al. 1994; Khanam et al. 2000; Raoufa et al. 2008). The major attraction of initiating and establishing suspension cultures is their property to easily adapt the submerged growth conditions and high potential to subsequent large-scale

Table 4.1 Some of the examples of tropane alkaloid production using plant in vitro cultures

In vitro culture	Plant	Family	Metabolite	References
Shoot cultures	*Erythroxylum coca* var. *coca*	Erythroxylaceae	Cocaine	Lydon et al. (1993)
Shoot cultures	*Scopolia parviflora*	Solanaceae	Hyoscyamine, Scopolamine	Kang et al. (2004)
Callus cultures	*Erythroxylum coca*	Erythroxylaceae	Cocaine	Docimo et al. (2015)
Suspension cultures	*Datura innoxia*	Solanaceae	Scopolamine	Gontier et al. (1994)
Root culture	*Atropa belladonna*	Solanaceae	Calystegines	Dräger et al. (1994)
Hairy root cultures (HRCs)	*Duboisia leichhardtii* F.	Solanaceae	Scopolamine	Mano et al. (1989)
HRCs	*Przewalskia tangutica*	Solanaceae	Hyoscyamine, scopolamine	Lan and Quan (2013)
HRCs	*Datura tatula* L., *D. stramonium*, *D. innoxia*	Solanaceae	Hyoscyamine	Harfi et al. (2018)
HRCs	*Scopolia lurida*	Solanaceae	Hyoscyamine	Zhao et al. (2017)
HRCs	*Datura metel*	Solanaceae	Atropine	Shakeran et al. (2015)
HRCs	*Hyoscyamus reticulates*, *H. pusillus*	Solanaceae	Hyoscyamine, scopolamine	Hedayati et al. (2020)
HRCs	*Anisodus acutangulus*	Solanaceae	Hyoscyamine, scopolamine, and anisodine	Kai et al. (2012a, b, c)
HRCs	*Anisodus luridus*	Solanaceae	Hyoscyamine, scopolamine	Qin et al. (2014)

culture in a variety of bioreactor. *H. muticus* suspension cultures exhibited rapid increase in biomass concomitant with accumulation of alkaloids when cultured in stirred tank and plastic-lined vessel bioreactors. The ability to produce callus and accumulation of alkaloids varies with the explants and associated growth conditions. For example, in *D. metel* callus cultures obtained from leaf, stem, and roots, the scopolamine content varied with the type of explant. Hyoscyamine was accumulated in root-borne callus only in concentration of 0.2 mg/g dry biomass (Raoufa et al. 2008). It has been observed that the overall accumulated alkaloid content in callus and cell suspension cultures is low; therefore, efforts to enhance the overall yield have also been made. In a study, immobilization of cell suspension of *D. innoxia* in Ca-alginate resulted in almost sixfold increase of both hyoscyamine and scopolamine contents (Gontier et al. 1994). Precursor feeding such as exogenous supplementation of tropic acid, tropinone, and tropanol as precursors of TA in *A. belladonna* suspensions led to increased synthesis of hyoscyamine and scopolamine (Simola et al. 1990).

In an attempt to search other production systems for TA production, shoot cultures of *H. albus*, *Duboisia leichhardtii*, *D. myoporoides*, *D. metel*, and *Scopolia parviflora* were established (Khanam et al. 2000; Kang et al. 2004). Although, negligible amount of hyoscyamine and scopolamine was observed in shoot cultures yet, in *D. leichhardtii* shoot cultures upon rooting and increase of root biomass, the hyoscyamine and scopolamine content were recorded 0.36% and 0.30% of shoot DW, respectively. The experiment strongly evidenced the importance of root organs in the biosynthesis of TAs (Georgiev et al. 2013).

Adventitious root cultures of TA-producing plants have also been explored for TA production (Maldonado-Mendoza et al. 1992; Kawamura et al. 1996; Min et al. 2007, 2009; Hong et al. 2010). Adventitious root culture of *H. niger* L. was cultivated in shake flasks, stirred tank, and mist bioreactor. No significant difference was observed in any of the culture system and yield obtained in a range between 0.05 and 0.075% of DW for hyoscyamine and 0.32% and 0.37% of DW for scopolamine. In contrast to this, biomass growth was found to decrease in both reactors as compared to shake flasks (Woo et al. 1996). Adventitious roots of *S. parviflora* and *A. belladonna* have also been grown in a small bubble column bioreactor (Min et al. 2007) and large-scale stirred tank bioreactor (Kawamura et al. 1996) to exhibit proportionate production of scopolamine only with that of biomass. To sum up, the prolonged culture duration coupled with the requirement of additional phytohormone supplementation and low yields makes adventitious root systems less attractive. Successful establishment of HRCs for TA production is reported in a number of TA-producing plants. Majority of plants belong to the family Solanaceae which include genera like *Atropa*, *Anisodus*, *Brugmansia*, *Datura*, *Duboisia*, *Hyoscyamus*, and *Scopolia* (Table 4.1).

4.3 Approaches Toward Higher TA Production

Though the production of TAs through in vitro interventions is reported to be very successful, however with the understanding of biology behind production, the idea of yield enhancement got attention. This can be done either by providing precursor to efficiently use SM pathways (precursor feeding), by engineering the pathway to either make it more stronger by overexpressing genes associated with rate-limiting steps, or by suppressing any of the diversion of steps to avoid undesired loss (metabolic engineering) (Srivastava et al. 2017a). Additionally, with the increased biomass, the desired SM metabolite production can also be achieved proportionally (Bioreactor Up-scaling). Few of the relevant examples under such categories are as follows:

4.3.1 Media Optimization and Precursor Feeding

It has been observed that changes in the composition of culture media influence SM production, and therefore this approach may be used to enhance metabolite production, including TAs. The effect of nitrate and ammonium concentrations was reported to have a strong influence on the scopolamine/hyoscyamine (S/H) ratio. In a study, it was reported that with increase in nitrate or ammonium concentrations, the ratio of S/H also increased by two to threefold in *A. belladonna* HRCs (Bensaddek et al. 2001). In another study, culture medium rich in or deprived with calcium, magnesium, and iron was used for *H. niger* roots. The result revealed that an increase in calcium in the medium unidirectionally decreases hyoscyamine while increasing scopolamine production. Additionally, alterations with exogenous magnesium and iron contents in the medium resulted in different changes in hyoscyamine and scopolamine concentrations (Pudersell et al. 2012). Similarly, in suspension cultures of *D. stramonium*, the effect of different concentrations of glucose, sucrose, $CaCl_2$, KNO_3, and NH_4NO_3 along with varying temperature was also tested (Iranbakhsh et al. 2007).

The early investigation of precursor feeding involves the administration of radiolabeled molecules to identify their fate in the biosynthesis of metabolite of interest. However, it has been observed later that though SM biosynthetic pathway enzymes are functional, still there is low production of metabolite of interest. Though in such cases, there is adequate enzymatic activity and chemical ambience for the production of metabolite of interest, the productivity is not satisfactory, due to paucity of its near or distant precursors (Srivastava et al. 2017b). To overcome this problem, the exogenous supplementation of pathway precursors is also suggested, with genesis that more raw supply could promote more outcomes (Table 4.2). This strategy in tissue culture is known as "precursor feeding" (Namdeo et al. 2007), which was explored in many SMs such as withanolide in *Withania somnifera* (Sivanandhan et al. 2014), indole alkaloids in *Catharanthus roseus* (Moreno et al. 1993), isoflavones in *Psoralea corylifolia* L. (Shinde et al. 2009), silymarin in *Silybum marianum* (Gharechahi et al. 2013), reserpine in *Rauwolfia serpentina* (Panwar and Guru 2015), etc. Precursor feeding is also considered an important choice to enhance the TA production in several genera. The increased production of scopolamine in the callus cultures of *Scopolia parviflora* was observed after supplementation of tropic acid in the culture medium (Tabata et al. 1972). Feeding of hyoscyamine and 6-hydroxy hyoscyamine in *Anisodus tanguiticus* suspension cultures resulted into scopolamine accumulation (Di et al. 1987). In the root cultures of *Duboisia leichhardtii*, the feeding of ^{14}C-labeled phenylalanine and tropic acid resulted into incorporation of radioactivity in the tropane alkaloids (Kitamura et al. 1992). The supplementation of TA precursor L-phenylalanine or DL-β-phenyllactic acid in *Datura innoxia* transformed roots has demonstrated higher hyoscyamine production (Boitel-Conti et al. 2000). In an attempt, the tobacco cell cultures carrying *HmH6H* (*H. muticus H6H*) derived from hairy roots, when administered with hyoscyamine, produce scopolamine (Moyano et al. 2007). Feeding of hyoscyamine and 6-hydroxy hyoscyamine in non-TA-producing *Anisodus tanguiticus* suspension cultures resulted into scopolamine accumulation (Di et al. 1987). The production of nortropane alkaloid calystegine was achieved in *Calystegia sepium*, after [^{15}N]tropinone was administered to root cultures (Scholl et al. 2003). Currently, this technique in combination with other yield enhancement strategies such as

Table 4.2 Biotransformation studies of in vitro cultures of tropane alkaloid-bearing plants

Plant system	In vitro cultures	Family	Substrate	Biotransformed product	Biotransformation strategies	References
Anisodus tanguticus	Plant cell suspension (PCS)	Solanaceae	D, L-hyoscyamine	6-Hydroxy hyoscyamine	Precursor feeding	Ke-Di et al. (1987)
Anisodus tanguticus	PCS	Solanaceae	6-Hydroxy hyoscyamine	Scopolamine	Precursor feeding	Chao and Lan (1989)
Datura fastuosa L.	PCS	Solanaceae	Hydroquinone	Arbutin	Exogenous molecule	Kittipongpatana et al. (2007)
Datura innoxia	PCS	Solanaceae	Tropic acid (RS)	Glucose ester	Precursor feeding	Tabacum (1989)
Datura innoxia	PCS	Solanaceae	Hydroquinone	Arbutin	Exogenous molecule	Suzuki et al. (1987)
Datura stramonium	PCS	Solanaceae	p-hydroxy benzaldehyde	Gastrodin	Exogenous molecule	Gong et al. (2006)
Hyoscyamus niger	PCS/ Root cultures	Solanaceae	Hyoscyamine	Scopolamine	Precursor feeding	Hashimoto and Yamada (1983)
Anisodus tanguticus	Hairy root cultures (HRCs)	Solanaceae	Dehydroepiandrosterone (DHEA)	Androst-4-ene-3,17-dione, 6 α-hydroxy androst-4-ene-3, 17-dione, 6 α-hydroxy androst-4-ene-3, 17-dione, androst-4-ene-3,6,17-trione, 17 β-hydroxy androst-4-ene-3-one	Exogenous molecule	Liu et al. (2004)
Atropa belladonna	HRCs	Solanaceae	Hyoscyamine	Scopolamine	Co-culture techniques	Subroto et al. (1996)
Atropa belladonna	HRCs	Solanaceae	Betuligenol	Raspberry ketone and betuloside	Exogenous molecule	Srivastava et al. (2013)
Atropa belladonna	HRCs	Solanaceae	3,4,5-Trimethoxy benzaldehyde	3,4,5-Trimethoxy benzoic acid and 3,4,5-Trimethoxy benzyl alcohol	Exogenous molecule	Srivastava et al. (2012)

(continued)

Table 4.2 (continued)

Plant system	In vitro cultures	Family	Substrate	Biotransformed product	Biotransformation strategies	References
Atropa belladonna	HRCs	Solanaceae	Artemisinin	3-α-Hydroxy-1-deoxyartemisinin and 4-hydroxy-9,10-dimethyloctahydrofuro-(3,2-i)-isochromen-11(4*H*)-one	Exogenous molecule	Pandey et al. (2015)
Brugmansia candida	HRCs	Solanaceae	Hydroquinone	Arbutin	Exogenous molecule	Casas et al. (1998)
Datura tatula L.	HRCs	Solanaceae	p-hydroxybenzyl alcohol	p-hydroxy-methyl phenol-β-D glucoside (Gastrodin)	Exogenous molecule	Peng et al. (2008)

metabolic engineering, elicitation, and bioreactor has been demonstrated to provide better results.

4.3.2 Elicitation

It is the process of induced and/or enhanced biosynthesis of SMs by supplementing with physical, chemical, or biological elicitor in the plant system (Namdeo 2007). After analysis of native production profile in medicinal plants, this is one of the most explored research areas. The elicitation studies have also been well studied in diverse in vitro system such as root culture, callus and cell suspension cultures, and hairy root cultures (Goel et al. 2011). In general, the HRC-based elicitation studies have demonstrated production of various highly active phytotherapeutics such as reserpine, scopolamine, hyoscyamine, podophyllotoxin, andrographolide, and tanshinone (Srivastava et al. 2017a).

In TA-bearing plants, this technique has been utilized in a number of plant system such as *Atropa*, *Hyoscyamus*, *Datura*, *Scopolia, Anisodus* etc. The *Datura stramonium* HRCs subjected to methyl jasmonate (MeJA), fungal elicitor, and oligogalacturonide demonstrated TA (littorine, hyoscyamine, and scopolamine) accumulation that ranges in the order MeJa > fungal elicitor > oligogalacturonide (Zabetakis et al. 1999). The *Datura* sp., viz., *D. stramonium*, *D. tatula*, and *D. innoxia* (cultivated for their hyoscyamine content), were raised and analyzed for hyoscyamine content after salicylic acid (SA) and acetyl salicylic acid (ASA) treatment. The study demonstrated increase hyoscyamine production, under optimal concentration (0.1 mM) for both elicitors (AS and ASA). In their study, the highest content was observed in *D. tatula* HRCs treated with SA (Harfi et al. 2018). Jung et al. (2003) have reported higher scopolamine production in HRCs of *Scopolia parviflora*, using bacterial elicitors. The HRCs of *Anisodus acutangulus* demonstrated the accumulation of tropane alkaloids up to 1.51, 1.13, and 1.08 times after 24 h posttreatment of ethanol, methyl jasmonate, and Ag^{+}; however, with salicylic acid, the average content of TAs was found to decrease. Further, the elicitation effect on upregulation of TA biosynthesis genes such as hyoscyamine-6β-hydroxylase (*AaH6H*) by ethanol and putrescine N-methyltransferase I (*AaPMT1*) by Ag^{+} and tropinone reductase I (AaTR1) by methyl jasmonate was also observed (Kai et al. 2012a, b, c). The treatment of colchicine and UV-B has reported to show the elicitation effect on hyoscyamine and scopolamine content in *H. reticulata* HRCs (Zeynali et al. 2016). In *H. niger* HRCs, MeJa-mediated scopolamine-elicited productivity was also observed (Jaremicz et al. 2014).

4.3.3 Metabolic Engineering

The metabolic engineering as the name itself justifies is a genetic engineering to regulate metabolite content. In plants, to enhance the content of desired metabolites, the pathway can be engineered at pathway level (engineering of genes associated with metabolic pathway), regulatory level (engineering of transcription factor that can regulate the expression of pathway genes), or combination of both (Srivastava et al. 2017a). Besides, it has also been observed that the metabolic pathway is not always linear and many diversions do exist, which reduces the overall metabolite productivity. To address this, the diverting pathway can be suppressed by inhibiting the expression of targeted genes (through RNAi). All these strategies are very successful in tailoring SM metabolic pathways in plants and lead to enhancement of many therapeutically important secondary metabolites, viz., camptothecin, ajmaline, codeine, reserpine, silymarin, paclitaxel, withanolides, etc. (Mehrotra et al. 2010; Srivastava et al. 2017a).

Majority of engineering is done on important and rate-limiting pathway genes including hyoscyamine-6β-hydroxylase (H6H; EC 1.14.11.11). The enzyme catalyzes oxidative reactions of the biosynthetic pathway that results in the conversion of hyoscyamine to scopolamine, a more pharmaceutically valuable tropane alkaloid. The *H. niger H6H* gene under the control of cauliflower mosaic virus 35S promoter was introduced to *A. belladonna* through a binary vector system using *A. rhizogenes* (Hashimoto et al. 1993). Such engineered *A. belladonna* hairy roots exhibited increased fivefold higher concentrations of scopolamine in comparison with wild-type hairy roots. In a combinatorial effort, where gene engineering was combined with precursor feeding, the *N. tabacum* and *H. muticus* hairy roots were studied for the production of scopolamine and other tropane and nicotine alkaloids after feeding the cultures with exogenous hyoscyamine. These roots of non-hyoscyamine-producing *N. tabacum* and hyoscyamine-producing *H. muticus* were transformed with CaMV 35S promoter-controlled *H6H* gene from *H. niger*. The high scopolamine in *H. muticus* and *N. tabacum* hairy roots carrying the 35S-*H6H* transgene revealed an efficient uptake of hyoscyamine from the culture medium and a higher rate of bioconversion of hyoscyamine to scopolamine.

The improvement of anisodine content was demonstrated in transgenic *Anisodus acutangulus* HRCs, by overexpression of branch-controlling enzyme tropinone reductase I (*AaTRI*) and the downstream rate-limiting enzyme hyoscyamine-6-β-hydroxylase (*AaH6H*) (Kai et al. 2012a, b, c). The overexpression of tropinone reductase (TRI and TRII) in *A. belladonna* root lines demonstrated enhanced production of enzyme products, tropine or pseudotropine, which leads to significant modulation of tropine-derived alkaloid and pseudotropine-derived alkaloid, respectively (Richter et al. 2005). Overexpression of *pmt* (putrescine N-methyltransferase) and *h6h* (hyoscyamine 6β-hydroxylase) genes resulted in increased scopolamine accumulation in transgenic plants of *A. belladonna* generated from hairy roots (Liu et al. 2010). Enhancement of tropane alkaloid production in *A. belladonna* hairy root cultures was also observed after overexpression of *pmt* and *h6h* genes (Yang et al. 2011). Metabolic engineering by co-expression of the putrescine N-methyl transferase (*SpPMT*) and hyoscyamine-6β hydroxylase (*SpH6H*) in *Scopolia parviflora* HRCs revealed enhanced hyoscyamine and scopolamine production (Kang et al. 2011). The overexpression of *PMT* in *Datura metel* and *H. muticus* HRCs suggested the improvement of both hyoscyamine and scopolamine and only hyoscyamine, respectively. Further, beyond HRCs, Moyano et al. (2007), while working on bioconversion of hyoscyamine and scopolamine in *HmH6H*-expressing tobacco cell line, have demonstrated increased scopolamine accumulation in 5 L-turbine stirred tank bioreactor.

4.4 Allied Applications of Tropane Alkaloid-Bearing In Vitro Cultures

4.4.1 Biotransformation

Biotransformation is defined as structural and/or conformational alteration of a chemical moiety by enzymatic militia of biological system, thereby providing green route for chemical derivatization. It offers perfect methodology to create structural changes in complex molecules, though the specificity of reactions cannot be determined in advance and needs initial experimentation. Though biotransformation is a fundamental feature of any life form and cumulatively responsible for the metabolism, yet its application was realized only when methodology to culture life came into existence. Currently, its application is frequently utilized to develop derivatives of naturally occurring phytomolecules by enzymatic machinery of microbe, animal, and plant systems (Srivastava et al. 2017b; Table 4.2).

Among plant in vitro cultures, mostly biotransformation studies were performed with cell suspension and hairy root cultures (Giri et al. 2001; Srivastava et al. 2017b). Further, this application in in vitro systems is independent of seasonal and pathological constraints and provides solutions for sustainable use of resources. These cultures offer many biotransformation reactions including hydroxylation, glycosylation, oxidoreduction, hydrolysis, etc. (Giri et al. 2001; Banerjee et al. 2012; Srivastava 2015; Srivastava et al. 2017b). Further, three types of biotransformation can be easily recognized based on culture method and substrate choice, viz., precursor feeding, co-culture, and non-specific/exogenous molecule biotransformation (Srivastava et al. 2017b).

Among tropane alkaloid-bearing plants, all these strategies were explored; however, the participation of non-specific/exogenous molecule biotransformation is very high. The precursor feeding as described earlier was considered as an effective approach toward elucidation of TA biosynthetic pathway or to enhance the metabolic flux (supply of early-step metabolite). Here, both the near (hyoscyamine toward scopolamine production) or distant precursor (phenylalanine toward scopolamine and tropinone toward calystegine production) have been utilized.

In co-culture practice, two complementary systems in single culture matrix (media) provide solutions for desired metabolite production. Here, one culture system provides substrate for the other system to biotransform, and there is no requirement for exogenous administration (Subroto et al. 1996). Subroto et al. (1996) have established the co-culture system comprised of *A. belladonna* shooty teratomas and HRCs (root-shoot coculture studies) in the same hormone-free medium, which demonstrated significant conversion of hyoscyamine to scopolamine as compared to individual root and shoot cultures. The suggested explanation revealed the sequence of event like hyoscyamine production in HRCs, its translocation through culture medium to shooty teratomas, and finally its conversion by shooty teratomas to scopolamine.

Out of the three approaches stated, the exogenous molecule biotransformation has been explored to a greater extent involving both cell suspension and HRCs (Giri et al. 2001; Banerjee et al. 2012; Srivastava et al. 2017b). Lucero et al. (1999) have reported the potentiality of *Datura innoxia* cell suspension to transform TNT into ADNT via nitroreduction. The bioconversion of artemisinin to deoxyartemisinin was reported in *Withania somnifera* (Sabir et al. 2010). Sabir et al. (2011) have also reported the biotransformation of withanolides (withanolide A, withaferin A, and withanone) using *W. somnifera* cell suspension, which demonstrated interconversion of withanolides and production of some unknown compounds.

Here also the number of publications achieved in cell suspension-based biotransformation has been overshadowed by HRC-based biotransformation in recent times, possibly due to HRC attributes, viz., genetic and biochemical stability and thus higher reproducibility, organ (differentiated structure) feature, etc. Among HRCs, *Atropa belladonna*, *Hyoscyamus muticus*, *Datura tatula*, *Brugmansia candida*, etc. were explored for biotransformation studies (Banerjee et al. 2012). The production of arbutin using hydroquinone was achieved with *B. candida* HRCs (Casas et al. 1998). Srivastava et al., while working on *A. belladonna* HRCs, reported biotransformation of 3,4,5-trimethoxy carbonyl compounds and betuligenol (Srivastava et al. 2012, 2013; Fig. 4.2). The transformations studies with betuligenol substrate lead to production of value molecules, viz., betuloside and raspberry ketone

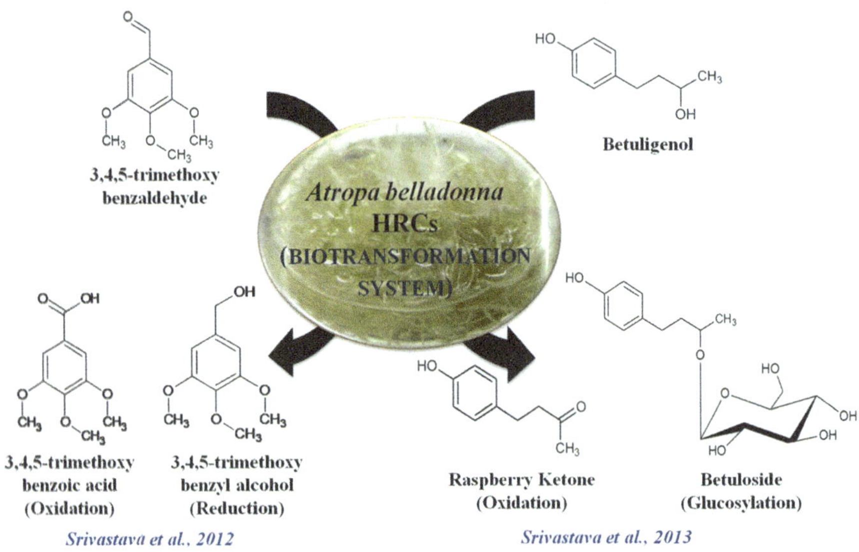

Fig. 4.2 Demonstration of few instances of *A. belladonna* HRC-mediated biotransformation studies

(Srivastava et al. 2013). Similarly the production of raspberry ketone through biotransformation in *H. muticus* was also reported (Hakkinen et al. 2015). Pandey et al. (2015) have utilized the *A. belladonna* and *H. muticus* HRCs for biotransformation of artemisinin into 3-α-hydroxy-1-deoxyartemisinin and 4-hydroxy-9,10-dimethyloctahydrofuro-(3,2-i)-isochromen-11(4*H*)-one; however, the conversion efficiency for these products differs among tested HRCs. In *D. tatula* HRCs, the biotransformation of p-hydroxybenzyl alcohol to gastrodin was reported (Peng et al. 2008).

4.4.2 Phytoremediation

Phytoremediation is a method of utilizing plants as remedy to clean up and restore the environment, which is a cheap, solar-driven, and eco-friendly approach. The exploration of plant cultures' ability to stimulate enzyme-based degradation, tolerance, and hyper-accumulation of pollutants in the solution has been noticed. During their metabolism, plants have the ability to utilize pollutants (organic or inorganic) and can remove them or transform them to less harmful derivatives (Mehrotra and Srivastava 2017). Phytoremediation study is also reported from in vitro cultures of tropane alkaloid-bearing plants for remediation of pollutants (Doran 2009; Mazaheri and Piri 2015). The complete clearance of TNT (2,4,6-trinitrotoluene) and ADNT (aminodinitrotoluenes) metabolites within 12 h was reported in *Datura innoxia* cell suspension (Locero et al. 1999). The cell suspension of *Datura stramonium* demonstrated to metabolize the pesticide 4[U^{-14}C]nitrophenol (Schmidt et al. 1997) and herbicide glufosinate-ammonium (Muller et al. 2001). Gonzalez et al. (2006) have reported the removal of phenol from water by tomato HRCs, involving peroxidase activity. In a study with *Atropa belladonna* HRCs, the P450 2E1 transgenic *A. belladonna* hairy root cultures with trichloroethylene (TCE) exhibited increased content of TCE metabolites (chloral and trichloroethanol) as compared to control (Banerjee et al. 2002). A patent on utility of *A. belladonna* HRCs for decontamination of medium containing polychlorinated biphenyls (PCBs) or dioxins was also reported (Morita et al. 2001). Recently, *A. belladonna* HRCs were also shown to remove phenol from wastewaters (Mazaheri and Piri 2015).

4.4.3 Molecular Farming

The utilization of whole plants or their cultured cell/tissue in vitro for the production of valuable recombinant proteins is called as plant molecular farming. It provides an economic alternative to conventional production system utilizing cultivation of microbial or mammalian cells in large-scale bioreactors (Schillberg et al. 2013). Ma et al. (2003) have compared bacteria, yeast, mammalian cell culture, transgenic animals, plant cell cultures, and transgenic plants, for recombinant human pharmaceutical proteins. The review suggested that plant-based systems have advantages like low cost, low risk of contamination, and at the same time high product quality. The first recombinant human pharmaceutical protein (human serum albumin) was reported by Sijmons et al. (1990) in tobacco suspension cultures. Since then, many antibodies, cytokines, enzymes, vaccines, etc. have been produced using plant in vitro cultures (Schillberg et al. 2013). Among TA-bearing plant, the molecular farming has also been attempted particularly in potato, tomato, and belladonna (Banerjee et al. 2002; Obembe et al. 2011; Singh et al. 2015). In in vitro cultures, such studies are mainly through HRCs. Banerjee et al. (2002) have reported the expression of P450 2E1, an important mammalian liver enzyme in *A. belladonna* HRCs. Singh et al. (2015) have reported the engineering of tomato hairy roots for expression of recombinant protein containing a fusion of rabies glycoprotein and ricin toxin B chain (rgp-rtxB) antigen.

4.5 Conclusion

The advancement in biotechnology undoubtedly provides ways to explore TA production. Many successful attempts for in vitro culture establishment and their analysis for TA metabolites are reported. This involves contribution of cell (callus and suspension) and organ (shoot and root) cultures and associated production. In addition to this, the development of *A. rhizogenes*-mediated HRCs have emerged as an effective culture option to address the issue in the most economical way. Along with this, the development of novel approaches for further enhancement of TAs, viz., media optimization, precursor feeding, elicitation, metabolic engineering, ploidy alteration, and bioreactor upscaling, further boosts the metabolite contents. However, the exploration under these approaches still needs efforts as the progress being made is limited at least for ploidy and bioreactor-based investigations, concerning TAs. Further, apart from being successful, the study mostly involves production toward only few metabolites (particularly hyoscyamine and scopolamine); however, attempts are also required to explore this technology for other therapeutic and commercially important TAs. The in vitro systems have also demonstrated a meaningful transition from "just metabolite production system" to "system for novel applications." Under such explorations, the contribution of biotransformation studies has proven their existence very effectively in TA-bearing plants (Giri et al. 2001; Srivastava et al. 2017a, b), followed by considerable progress in phytoremediation (Namdeo et al. 2007). However, the progress made in in vitro culture-based molecular farming is limited in such plants and needs further exploration.

Acknowledgments VS, SL, and AC acknowledge the Central University of Jammu for providing work facility.

References

Banerjee S, Shang TQ, Wilson AM, Moore AL, Strand SE, Gordon MP, Lafferty Doty S (2002) Expression of functional mammalian P450 2E1 in hairy root cultures. Biotechnol Bioeng 77 (4):462–466

Banerjee S, Singh S, Rahman LU (2012) Biotransformation studies using hairy root cultures—a review. Biotechnol Adv 30(3):461–468

Bensaddek L, Gillet F, Saucedo JEN, Fliniaux MA (2001) The effect of nitrate and ammonium concentrations on growth and alkaloid accumulation of Atropa belladonna hairy roots. J Biotechnol 85(1):35–40

Boitel-Conti M, Laberche JC, Lanoue A, Ducrocq C, Sangwan-Norreel BS (2000) Influence of feeding precursors on tropane alkaloid production during an abiotic stress in Datura innoxia transformed roots. Plant Cell Tiss Org Cult 60(2):131–137

Canter PH, Thomas H, Ernst E (2005) Bringing medicinal plants into cultivation: opportunities and challenges for biotechnology. Trends Biotechnol 23(4):180–185

Casas DA, Pitta-Alvarez SI, Giulietti AM (1998) Biotransformation of hydroquinone by hairy roots of Brugmansia candida and effect of sugars and free-radical scavengers. Appl Biochem Biotechnol 69:127–136

Chandran H, Meena M, Barupal T (2020) Sharma K plant tissue culture as a perpetual source for production of industrially important bioactive compounds. Biotechnol Rep 26:e00450

Chao M, Lan L (1989) The biotransformation of 6-hydroxyhyoscyamine to scopolamine by suspension and immobilized cells of Anisodus tanguticus. Planta Med 55(04):391–392

Docimo T, Davis AJ, Luck K, Fellenberg C, Reichelt M, Phillips M, Gershenzon J, D'Auria JC (2015) Influence of medium and elicitors on the production of cocaine, amino acids and phytohormones by Erythroxylum coca calli. Plant Cell Tissue Organ Culture (PCTOC) 120 (3):1061–1075

Doran PM (2009) Application of plant tissue cultures in phytoremediation research: incentives and limitations. Biotechnol Bioeng 103(1):60–76

Dräger B, Funck C, Höhler A, Mrachatz G, Nahrstedt A, Portsteffen A, Schaal A, Schmidt R (1994) Calystegines as a new group of tropane alkaloids in Solanaceae. Plant Cell Tiss Org Cult 38 (2–3):235–240

Espinosa-Leal CA, Puente-Garza CA, García-Lara S (2018) In vitro plant tissue culture: means for production of biological active compounds. Planta 248(1):1–18. https://doi.org/10.1007/s00425-018-2910-1

Georgiev V, Marchev A, Berkov S, Pavlov A (2013) Plant in vitro systems as sources of tropane alkaloids. Nat Prod 5:173–211. https://doi.org/10.1007/978-3-642-22144-6_6

Gharechahi J, Khalili M, Hasanloo T, Salekdeh GH (2013) An integrated proteomic approach to decipher the effect of methyl jasmonate elicitation on the proteome of Silybummarianum L. hairy roots. Plant Physiol Biochem 70:115–122

Giri A, Dhingra V, Giri CC, Singh A, Ward OP, Narasu ML (2001) Biotransformations using plant cells, organ cultures and enzyme systems: current trends and future prospects. Biotechnol Adv 19(3):175–199

Goel MK, Mehrotra S, Kukreja AK (2011) Elicitor-induced cellular and molecular events are responsible for productivity enhancement in hairy root cultures: an insight study. Appl Biochem Biotechnol 165:1342–1355

Gong JS, Ma WP, Pu JX, Xu SG, Zheng SQ, Xiao CJ (2006) The production of gastrodin through biotransformation of p-hydroxybenzaldehyde by cell suspension culture of Datura stramonium. Acta Pharm Sin 41(10):963

Gontier E, Sangwan BS, Barbotin JN (1994) Effects of calcium, alginate and calcium-alginate immobilization on growth and tropane alkaloid levels of a stable suspension cell line of *Datura innoxia*. Plant Cell Rep 13:533–536

González PS, Capozucca CE, Tigier HA, Milrad SR, Agostini E (2006) Phytoremediation of phenol from wastewater, by peroxidases of tomato hairy root cultures. Enzyme Microb Technol 39 (4):647–653

Gutierrez-Valdes N, Häkkinen ST, Lemasson C, Guillet M, Oksman-Caldentey KM, Ritala A, Cardon F (2020) Hairy root cultures—a versatile tool with multiple applications. Front Plant Sci 11:33

Häkkinen ST, Seppänen-Laakso T, Oksman-Caldentey KM, Rischer H (2015) Bioconversion to raspberry ketone is achieved by several non-related plant cell cultures. Front Plant Sci 6:1035

Harfi B, Khelifi L, Khelifi-Slaoui M, Assaf-Ducrocq C, Gontier E (2018) Tropane alkaloids GC/MS analysis and low dose elicitors' effects on hyoscyamine biosynthetic pathway in hairy roots of Algerian Datura species. Sci Rep 8(1):1–8

Hashimoto T, Yamada Y (1983) Scopolamine production in suspension cultures and redifferentiated roots of Hyoscyamus niger. Planta Med 47(04):195–199

Hashimoto T, Yun DJ, Yamada Y (1993) Production of tropane alkaloids in genetically engineered root cultures. Phytochemistry 32(3):713–718

Hedayati A, Hosseini B, Palazon J, Maleki R (2020) Improved tropane alkaloid production and changes in gene expression in hairy root cultures of two Hyoscyamus species elicited by silicon dioxide nanoparticles. Plant Physiol Biochem 155:416–428

Herouart D, Gontier E, Sangwan RS, Sangwan-Norreel BS (1991) Analysis of the potentialuse of androgenic Datura innoxia for the development of cell cultures producing highamounts of tropane alkaloids. J Exp Bot 42(8):1073–1076

Hong MLK, Bhatt A, Shuping N, Keng CL (2010) Optimization of root proliferation medium for Hyoscyamus niger L. J Med Plant Res 4(24):2710–2718

Hunter P (2011) A new era for plant biotechnology. A combination of pharmaceutical applications, technologies and safety measures could lead to new applications for genetically modified plants. EMBO Rep 12(6):504–507. https://doi.org/10.1038/embor.2011.91

Iranbakhsh AR, Oshagi MA, Ebadi M (2007) Growth and production optimization of tropane alkaloids in Datura stramonium cell suspension culture. Pak J Biol Sci 10:1236–1242

Jaremicz Z, Luczkiewicz M, Kokotkiewicz A, Krolicka A, Sowinski P (2014) Production of tropane alkaloids in Hyoscyamus niger (black henbane) hairy roots grown in bubble-column and spray bioreactors. Biotechnol Lett 36(4):843–853

Jung HY, Kang SM, Kang YM, Kang MJ, Yun DJ, Bahk JD, Yang JK, Choi MS (2003) Enhanced production of scopolamine by bacterial elicitors in adventitious hairy root cultures of Scopoliaparviflora. Enzyme Microb Technol 33(7):987–990

Kai G, Yang S, Zhang Y, Luo X, Fu X, Zhang A, Xiao J (2012a) Effects of different elicitors on yield of tropane alkaloids in hairy roots of Anisodusacutangulus. Mol Biol Rep 39 (2):1721–1729

Kai G, Yang S, Zhang Y, Luo X, Fu X, Zhang A, Xiao J (2012b) Effects of different elicitors on yield of tropane alkaloids in hairy roots of Anisodus acutangulus. Mol Biol Rep 39 (2):1721–1729

Kai G, Zhang A, Guo Y, Li L, Cui L, Luo X, Liu C, Xiao J (2012c) Enhancing the production of tropane alkaloids in transgenic *Anisodus acutangulus* hairy root cultures by over-expressing tropinonereductase I and hyoscyamine-6β-hydroxylase. Mol Bio Syst 8(11):2883–2890

Kang YM, Min JY, Moon HS, Karigar CS, Prasad DT, Lee CH, Choi MS (2004) Rapid in vitro adventitious shoot propagation of *Scopolia parviflora* through rhizome cultures for enhanced production of tropane alkaloids. Plant Cell Rep 23(3):128–133. https://doi.org/10.1007/s00299-004-0820-0

Kang YM, Park DJ, Min JY, Song HJ, Jeong MJ, Kim YD, Kang SM, Karigar CS, Choi MS (2011) Enhanced production of tropane alkaloids in transgenic Scopolia parviflora hairy root cultures over-expressing putrescine N-methyl transferase (PMT) and hyoscyamine-6β-hydroxylase (H6H). In Vitro Cell Dev Biol Plant 47(4):516–524

Kawamura M, Shigeoka T, Akita M, Kobayashi Y (1996) Newly developed apparatus for inoculating plant organs into large-scale fermentor. J Ferment Bioeng 82(6):618–619

Ke-Di C, Wei-Hua Z, Xin-Lan L, Chao M, Zai-Ming S, Dan-Hua Y (1987) Biotransformation of hyoscyamine by suspension cultures of Anisodustanguticus. Planta 53(02):211–213
Khanam N, Khoo C, Close R, Khan AG (2000) Organogenesis, differentiation and histolocalization of alkaloids in cultured tissues and organs of Duboisia myoporoides R. Br Ann Bot 86 (4):745–752. https://doi.org/10.1006/anbo.2000.124753
Kinsara AM, Seif El-Naser MM (1990) Growth and production of tropane alkaloids by cell suspension cultures of Hyoscyamus albus L. Qatar Univ Sci Bull 10:145–153
Kitamura Y, Taura A, Kajiya Y, Miura H (1992) Conversion of phenylalanine and tropic acid into tropane alkaloids by Duboisia leichhardtii root cultures. J Plant Physiol 140(2):141–146
Kittipongpatana N, Tangjai T, Srimanee A, Kittipongpatana OS (2007) Biotransformation of hydroquinone to arbutin by cell-suspension cultures of three Thai Solanaceous plants. CMU J Nat Sci 6:219–229
Lan X, Quan H (2013) Hairy root culture of Przewalskia tangutica for enhanced production of pharmaceutical tropane alkaloids. J Med Plant Res 4(14):1477–1481
Liu Y, Cheng KD, Zhu P, Feng WH, Meng C, Zhu HX, He HX, Ma XJ (2004) Biotransformation of dehydroepiandrosterone by hairy root cultures of Anisodus tanguticus. Acta pharmaceutica Sinica 39(6):445–448
Liu X, Yang C, Chen M, Li M, Liao Z, Tang K (2010) Promoting scopolamine accumulation in transgenic plants of Atropa belladonna generated from hairy roots with over expression of pmt and h6h gene. J Med Plant Res 4(17):1708–1713
Lucero ME, Mueller W, Hubstenberger J, Phillips GC, O'Connell MA (1999) Tolerance to nitrogenous explosives and metabolism of TNT by cell suspensions of Datura innoxia. In Vitro Cell Develop Biol Plant 35(6):480–486
Lydon J, Zimmerman RH, Fordham IM, Lusby WR (1993) Tissue culture and alkaloid production of Erthroxylum coca var. coca. Int J Geogr Inf Syst 2(1):3–14
Ma JK, Drake PM, Christou P (2003) The production of recombinant pharmaceutical proteins in plants. Nat Rev Genet 4(10):794–805
Maldonado-Mendoza IE, Del Ayora-Talavera TR, Loyola-Vargas VM (1992) Tropane alkaloid production in Datura stramonium root cultures. In Vitro Cell Dev Biol Plant 28(2):67–72
Mano Y, Ohkawa H, Yamada Y (1989) Production of tropane alkaloids by hairy root cultures of Duboisia leichhardtii transformed by agrobacterium rhizogenes. Plant Sci 59(2):191–201
Mazaheri H, Piri K (2015) Removal of phenol by *A. belladonna* L. hairy root. Int J Phytoremediation 17:1212–1219
Mehrotra S, Rahman LU, Kukreja AK (2010) An extensive case study of hairy-root cultures for enhanced secondary-metabolite production through metabolic-pathway engineering. Biotechnol Appl Biochem 56:161–172
Mehrotra S, Srivastava V, Ur Rahman L, Kukreja AK (2015) Hairy root biotechnology-indicative timeline to understand missing links and future outlook. Protoplasma 252(5):1189–1201
Mehrotra S, Srivastava V (2017) Hairy root in vitro systems: a suitable biological matrix for plant based remediation of environmental pollutants. In: Chemical pollution control with microorganisms. Nova Science Publishers, Hauppauge, pp 245–274
Min JY, Jung HY, Kang SM, Kim YD, Kang YM, Park DJ, Prasad DT, Choi MS (2007) Production of tropane alkaloids by small-scale bubble column bioreactor cultures of Scopoliaparviflora adventitious roots. Bioresour Technol 98(9):1748–1753. https://doi.org/10.1016/j.biortech.2006.07.03359
Min JY, Park DJ, Jeong MJ, Song HJ, Kang SM, Kang YM, Choi MS (2009) Selection of tropane alkaloids high-producing lines by single cell cloning of Hyoscyamus niger L. root cultures. J Korean For Soc 98(2):142–147
Moreno PRH, Van der Heijden R, Verpoorte R (1993) Effect of terpenoid precursor feeding and elicitation on formation of indole alkaloids in cell suspension cultures of Catharanthusroseus. Plant Cell Rep 12(12):702–705

Morita M, Yamazaki T, Kamiya T, Takano H, Fuse O, Manabe E, Maruta T, Taiheiyo Cement Corp (2001) Method of decontaminating medium containing polychlorinated biphenyls or dioxins. US Patent 6,303,844

Moyano E, Palazón J, Bonfill M, Osuna L, Cusidó RM, Oksman-Caldentey KM, Pinol MT (2007) Biotransformation of hyoscyamine into scopolamine in transgenic tobacco cell cultures. J Plant Physiol 164(4):521–524

Müller BP, Zumdick A, Schuphan I, Schmidt B (2001) Metabolism of the herbicide glufosinate-ammonium in plant cell cultures of transgenic (rhizomania-resistant) and non-transgenic sugarbeet (Beta vulgaris), carrot (Daucuscarota), purple foxglove (Digitalis purpurea) and thorn apple (Datura stramonium). Pest Manag Sci 57(1):46–56

Namdeo AG (2007) Plant cell elicitation for production of secondary metabolites: a review. Pharmacogn Rev 1(1):69–79

Namdeo AG, Jadhav TS, Rai PK, Gavali S, Mahadik KR (2007) Review article precursor feeding for enhanced production of secondary metabolites: a review. Pharmacognosy Rev 1:2

Obembe OO, Popoola JO, Leelavathi S, Reddy SV (2011) Advances in plant molecular farming. Biotechnol Adv 29(2):210–222

Oseni OM (2018) A review on plant tissue culture, a technique for propagation and conservation of endangered plant species. Int J Curr Microbiol App Sci 7:3778–3786. https://doi.org/10.20546/ijcmas.2018.707.438

Pandey P, Singh S, Tewari N, Srinivas KVNS, Shukla A, Gupta N, Vasudev PG, Khan F, Pal A, Bhakuni RS, Tandon S (2015) Hairy root mediated functional derivatization of artemisinin and their bioactivity analysis. J Mol Catal B Enzym 113:95–103

Panwar GS, Guru SK (2015) Stimulation of reserpine production in the whole plant culture of Rauwolfiaserpentina L. by elicitors and precursor feeding. J Plant Biochem Biotechnol 24 (1):49–55

Peng CX, Gong JS, Zhang XF, Zhang M, Zheng SQ (2008) Production of gastrodin through biotransformation of p-hydroxybenzyl alcohol using hairy root cultures of Datura tatula L. Afr J Biotechnol 7(3):7

Pudersell K, Vardja T, Vardja R, Matto V, Arak E, Raal A (2012) Inorganic ions in the medium modify tropane alkaloids and riboflavin output in *Hyoscyamus niger* root cultures. Pharmacogn Mag 8(29):73–77. https://doi.org/10.4103/0973-1296.93330

Qin B, Ma L, Wang Y, Chen M, Lan X, Wu N, Liao Z (2014) Effects of acetylsalicylic acid and UV-B on gene expression and tropane alkaloid biosynthesis in hairy root cultures of Anisodus luridus. Plant Cell Tissue Organ Culture (PCTOC) 117(3):483–490

Raoufa A, El-Wakil H, Gabal A, El-Said A, Khlifa D (2008) Production of scopolamine and hyoscyamine in calli and regenerate cultures of Datura metel (L.). J Appl Sci Res 4:1858–1866

Richter U, Rothe G, Fabian AK, Rahfeld B, Dräger B (2005) Overexpression of tropinonereductases alters alkaloid composition in Atropa belladonna root cultures. J Exp Bot 56(412):645–652

Sabir F, Kumar A, Tiwari P, Pathak N, Sangwan RS, Bhakuni RS, Sangwan NS (2010) Bioconversion of artemisinin to its nonperoxidic derivative deoxyartemisinin through suspension cultures of Withaniasomnifera Dunal. Z Naturforsch C 65(9–10):607–612

Sabir F, Sangwan RS, Singh J, Misra LN, Pathak N, Sangwan NS (2011) Biotransformation of withanolides by cell suspension cultures of Withaniasomnifera (Dunal). Plant Biotechnol Rep 5 (2):127–134

Schillberg S, Raven N, Fischer R, Twyman M, R. and Schiermeyer, A. (2013) Molecular farming of pharmaceutical proteins using plant suspension cell and tissue cultures. Curr Pharm Des 19 (31):5531–5542

Schmidt B, Breuer J, Thiede B, Schuphan I (1997) Metabolism of the pesticide metabolite 4-[U-14C] Nitrophenolin Daturastramonium (L.) cell suspension culture—A model for processing of xenobiotic carbohydrate conjugates and formation of nonextractable residues. Pestic Biochem Physiol 57(2):109–118

Scholl Y, Schneider B, Dräger B (2003) Biosynthesis of calystegines: 15N NMR and kinetics of formation in root cultures of Calystegiasepium. Phytochemistry 62(3):325–332

Shakeran Z, Keyhanfar M, Asghari G, Ghanadian M (2015) Improvement of atropine production by different biotic and abiotic elicitors in hairy root cultures of Datura metel. Turk J Biol 39 (1):111–118

Shinde AN, Malpathak N, Fulzele DP (2009) Enhanced production of phytoestrogenicisoflavones from hairy root cultures of Psoraleacorylifolia L. using elicitation and precursor feeding. Biotechnol Bioprocess Eng 14(3):288

Simola LK et al (1990) Feeding experiments using suspension cultures of Atropa belladonna; limiting steps in the biosynthesis of tropane alkaloids. Phytochemistry 29(11):3517–3523

Sijmons PC, Dekker BM, Schrammeijer B, Verwoerd TC, Van Den Elzen PJ, Hoekema A (1990) Production of correctly processed human serum albumin in transgenic plants. Biotechnology 8(3):217–221

Singh A, Srivastava S, Chouksey A, Panwar BS, Verma PC, Roy S, Singh PK, Saxena G, Tuli R (2015) Expression of rabies glycoprotein and ricin toxin B chain (RGP–RTB) fusion protein in tomato hairy roots: a step towards Oral vaccination for rabies. Mol Biotechnol 57(4):359–370

Sivanandhan G, Selvaraj N, Ganapathi A, Manickavasagam M (2014) Enhanced biosynthesis of withanolides by elicitation and precursor feeding in cell suspension culture of Withaniasomnifera (L.) Dunal in shake-flask culture and bioreactor. PLoS One 9(8):e104005

Skała E, Sitarek P (2020) Hairy roots as a source of tropane alkaloids. In: Srivastava V, Mehrotra S, Mishra S (eds) Hairy root cultures based applications. Rhizosphere biology. Springer, Singapore

Srivastava V (2015) Studies on hairy roots as bioreactors for normal and elicited biotransformation of exogenous molecules. PhD thesis, University of Lucknow, Lucknow

Srivastava V, Negi AS, Ajayakumar PV, Khan SA, Banerjee S (2012) Atropa belladonna hairy roots: orchestration of concurrent oxidation and reduction reactions for biotransformation of carbonyl compounds. Appl Biochem Biotechnol 166(6):1401–1408

Srivastava V, Kaur R, Chattopadhyay SK, Banerjee S (2013) Production of industrially important cosmaceutical and pharmaceutical derivatives of betuligenol by Atropa belladonna hairy root mediated biotransformation. Ind Crop Prod 44:171–175

Srivastava V, Mehrotra S, Verma PK (2017a) Biotechnological interventions for production of therapeutic secondary metabolites using hairy root cultures of medicinal plants. In: Current developments in biotechnology and bioengineering; crop modification, nutrition, and food production. Elsevier Science, Amsterdam, pp 259–282

Srivastava V, Mehrotra S, Mishra S (2017b) Biotransformation through hairy roots: perspectives, outcomes, and major challenges. In: Jha S (ed) Transgenesis and secondary metabolism. Springer, Cham, pp 1–24

Srivastava V, Mehrotra S, Mishra S (2018) Hairy roots: an effective tool of plant biotechnology. Springer, Cham

Srivastava V, Mehrotra S, Mishra S (2020) Hairy root cultures based applications. Springer, Cham

Subroto MA, Kwok KH, Hamill JD, Doran PM (1996) Coculture of genetically transformed roots and shoots for synthesis, translocation, and biotransformation of secondary metabolites. Biotechnol Bioeng 49(5):481–494

Suzuki T, Yoshioka T, Tabata M, Fujita Y (1987) Potential of Datura innoxia cell suspension cultures for glucosylating hydroquinone. Plant Cell Rep 6(4):275–278

Tabacum N (1989) Biotransformation of (RS)-tropic acid in suspension cultures. Phytochemistry 28(9):2333–2339

Tabata M, Yamamoto H, Hiraoka N, Konoshima M (1972) Organization and alkaloid production in tissue cultures of Scopoliaparviflora. Phytochemistry 11(3):949–955

Woo SH, Park JM, Yang JW (1996) Root culture using a mist culture system and estimation of scale-up feasibility. J Chem Technol Biotechnol 65(4):355–362

Yamada Y, Endo T (1984) Tropane alkaloid production in cultured cells of Duboisialeichhardtii. Plant Cell Rep 3(5):186–188

Yamada Y, Hashimoto T (1982) Production of tropane alkaloids in cultured cells of Hyoscyamus niger. Plant Cell Rep 1(3):101–103

Yang C, Chen M, Zeng L, Zhang L, Liu X, Lan X, Tang K, Liao Z (2011) Improvement of tropane alkaloids production in hairy root cultures of'Atropabelladonna'by overexpressing pmt and h6h genes. Plant Omics 4(1):29

Zabetakis I, Edwards R, O'Hagan D (1999) Elicitation of tropane alkaloid biosynthesis in transformed root cultures of Datura stramonium. Phytochemistry 50(1):53–56

Zeynali Z, Hosseini B, Rezaei E (2016) Effect of elicitation on antioxidant activity and production of tropane alkaloids in Hyoscyamus reticulatus hairy root cultures. Res J Pharm 3(3):43–53

Zhao K, Zeng J, Zhao T, Zhang H, Qiu F, Yang C, Zeng L, Liu X, Chen M, Lan X, Liao Z (2017) Enhancing tropane alkaloid production based on the functional identification of tropine-forming reductase in Scopolia lurida, a tibetan medicinal plant. Front Plant Sci 8:1745

Hairy Root Cultures and Plant Tropane Alkaloids: Production Matrix and Product in Biotechnological Perspective

5

Shital K. Sharma and Adarsh K. Agnihotri

Abstract

Plant tropane alkaloids are the secondary metabolites known for their immense pharmaceutical importance and have been a sole ingredient of various ancient and modern medicinal drug formulations. These drugs are meant for the treatment of a range of simple to serious ailments like muscle spasms, stomachaches, blood circulatory diseases, and chronic obstructive pulmonary diseases. Members of Solanaceae and other plant families like Erythroxylaceae, Convolvulaceae, and Brassicaceae are the natural repositories of tropane alkaloids and have been exploited globally for their biotechnological production. Hairy root cultures of the various plant genera belonging to these families have proven their worth as a suitable production system for these metabolites and are widely utilized for the biotechnological production of tropane alkaloids. The present chapter explores the hitherto and current status in advancement of using hairy root cultures of TA-producing plants for the biotechnological production of tropane alkaloid. The text focuses on various primary and additional advanced strategies for establishment, optimization, and enhancement of tropane alkaloid yields using hairy root cultures.

Keywords

Hairy roots · Solanaceae · Tropane alkaloids · Bioreactor

S. K. Sharma (✉) · A. K. Agnihotri
Bio-processing and Herbal Division, Mahatma Gandhi Institute for Rural Industrialization, Wardha, Maharashtra, India

V. Srivastava et al. (eds.), *Tropane Alkaloids*,
https://doi.org/10.1007/978-981-33-4535-5_5

Abbreviations

DW	Dry weight
FW	Fresh weight
H6H	Hyoscyamine 6-β-hydroxylase
HRCs	Hairy root cultures
MJ	Methyl jasmonate
MPUT	Methylputrescine
PMT	Putrescine N-methyltransferase
SA	Salicylic acid
SNP	Sodium nitroprusside
TAs	Tropane alkaloids
TRI	Tropine reductases

5.1 Introduction

Plants are the rich source of a wide range of bioactive compounds. These compounds are used as raw materials for various industries including pharmaceutical, flavor, dyes, cosmetics, and fragrance. In a broader view, alkaloids, terpenes, and derivatives of phenols are major groups of plant-derived natural bioactive compounds (Mehrotra et al. 2020). Alkaloids are known to be produced by plants, animals, and even microbes; however, a wide array of alkaloids, including tropane alkaloids, are produced by plants (Goyal 2013). Tropane alkaloids (TAs) are naturally occurring; significant, bioactive organic compounds occurred in plants all over the world and are known to mankind since ancient times primarily for their medicinal uses (Kohnen-Johannsen and Kayser 2019). Some of the tropane alkaloid-bearing plants like *Aropa*, *Datura*, *Hyoscyamus*, and *Duboisia* are known to humans since historic times due to their frequent medicinal uses. The compounds got their name from the Greek word *Atropos*, the name of the Goddess of life and death. This is because of the toxic and lethal property of these compounds. Basically, TAs are ornithine-derived compounds that transpire in many members of the families like Solanaceae, Erythroxylaceae, Convolvulaceae, Brassicaceae, and Euphorbiaceae and are characterized by their unique bicyclic tropane ring system (Kohnen-Johannsen and Kayser 2019). The occurrence of tropane skeletal structure with pyrrolidine and piperidine ring sharing nitrogen and two carbons is the basic structural property of these compounds. The IUPAC name of these compounds is 8-methyl-8-azabicyclo[3.2.1.]octane. Amino acid L-ornithine is the precursor compound for the biosynthesis of TAs. The common steps of chemical conversion and biosynthetic pathway for the formation of tropane skeleton from precursor amino acid are well documented in the literature [3, 4 and reference therein].

Naturally, in plant species, this group of alkaloids do not found free rather as part of esters (Hasan and Hasibul 2017). Esters of tropane alkaloids are the most common secondary metabolites of these plants. Tropane alkaloids are highly bioactive and of immense pharmaceutical value due to which they are globally used as active ingredient of a large number of commercially available drug formulations, for example, belladonna, used in the treatment of asthma, whooping cough, motion sickness, and Parkinson's disease (Paul and Datta 2011). Owing to such pharmaceutical significance, the huge commercial demand for plant TAs always remains unmet. Majority of plant TAs are directly harvested from source plant, leading to heavy loss of natural resources. The unsystematic management of the cultivation of genetic resources and harvesting from them has resulted in an unbalanced demand and supply ratio. Normally, these alkaloids synthesize and accumulate in underground parts of the native plant and take more than 3 years of field maturation of plant for ready to harvest root mass. This extended maturation phase combined with other climatic and pathological interruptions has led to an increasing interest to the other options for TA production. Rather sticking on aforementioned traditional unsystematic harvesting approach, the nonconventional options for the production of important plant TAs have attracted the scientific workers. The most talked about nonconventional method for the production of plants and plant-based secondary compounds is through biotechnological intervention (Ahmad et al. 2013; Gonçalves and Anabela 2018). Various in vitro strategies such as callus and suspension cultures, whole-plant cultures, and adventitious or hairy root cultures have been adopted for biotechnological exploration and production of plant TAs (Oksman-Caldentey and Arroo 2000; Georgiev et al. 2013). The success stories of these in vitro approaches for the production of TAs or any other plant secondary metabolite are not the focus of this article. However, involvement of genetically transformed hairy root cultures of TA-producing plants for their biotechnological production is being discussed in upcoming text.

5.2 Hairy Root Cultures (HRCs)

Among the different in vitro systems, hairy root cultures have proven their worth not only as a suitable biotechnological production system but also as an investigative platform for numerous biological studies (Mehrotra et al. 2020; Georgiev et al. 2013). Hairy root cultures are the specific type of in vitro cultures that are established by exploiting the unique natural phenomenon of horizontal gene transfer from the genus *Agrobacterium* to plants (Mehrotra et al. 2015). During this gene transfer process, a small DNA segment named transfer DNA or tDNA from bacterial root-inducing plasmid (Ri plasmid) is transferred and permanently integrated into the random site of host plant genome. A plethora of literature is available regarding these hairy root cultures and their applications that can satisfy the knowledge thrust of new comers to experts of this subject (Mehrotra et al. 2015; Gutierrez-Valdes et al. 2020; with reference therein). In a nut shell, hairy roots are emerged on the sites of plants where soilborne Gram-negative *Agrobacterium* invades the tissues due to the presence of wounds or other natural openings. The wounded tissues do release a specific phenolic compound acetosyringone and sugars, which induces the expression of several *chv* and *vir* genes of bacterial chromosome and Ri plasmid, respectively. This induction triggers the commencement of a phenomenon consisting of several synchronized processes of horizontal DNA transfer from bacterium to plant cell (Chilton et al. 1982). These include T-complex formation in bacterium, its further transportation to the plant cell, nuclei localization, and T-DNA integration. The molecular mechanisms behind various steps of this multipart and intricate inter-kingdom horizontal gene transfer are very well studied, and plenty of literature is available to understand this phenomenon from bacterial as well as host plant perspectives (Gelvin 2009; Georgiev et al. 2012). In general, two different tDNA fragments (TL-DNA and TR-DNA) from left and right sides of tDNA segment could be transferred independently into the plant cell. Only TL-DNA encodes the genes responsible for neoplastic disease symptom in terms of formation of hairlike root mass at infection sites of host plants. On the other hand, the TR-DNA codes for aux1, aux2, RolB TR, mas1, mas2, and ags genes that are responsible for opine and auxin biosynthesis in plants (Chandra 2012). The expression of *A. rhizogenes* oncogenes integrated in host genome along with tDNA segment can agitate the signal transduction pathway involved in auxin perception and biosynthesis in plant cells. This results in cumulative sensitivity of plant cells to endogenous phytohormones and thus leads to a fast-growing, highly branched mass of roots called as hairy roots by each transformed host plant cell (Gelvin 2009; Georgiev et al. 2012; Chandra 2012; Mehrotra et al. 2015, with references therein).

Besides the striking properties of genetic and biochemical stability, these hairy roots are able to grow and maintain for a remarkably long time if subcultured timely. Very often, they exhibit similar and/or higher levels of native and/or novel secondary metabolites of host plants. The utility of hairy roots for the production of plant-derived secondary metabolites does not require preparatory or elementary discussion. These in vitro systems have been explored globally for the secondary metabolite production purpose and have glorious approximately 50 year's history of being flawless biotechnological matrices for this purpose (Mehrotra et al. 2015). With reference to the biotechnological production of tropane alkaloids, hairy root cultures have shown their extensive utilization in exploring different strategies and standardization of different protocols at both bench-scale and large-scale bioreactor which will be discussed in ensuing text. Although hairy root cultures of members of the family Solanaceae hold the prime spot in this regard, the members of other TA-producing plant families and their hairy root cultures have equally garnered the attention of the scientific community (Kim et al. 2016; Gutierrez-Valdes et al. 2020). The upcoming text comprehensively focuses on the contribution of hairy root cultures for the biotechnological production of plant TAs. Further, the text also reflects the involvement of different TA-harboring plant families in evolving the hairy root culture systems as an impeccable and practically feasible approach for harnessing their naturally occurring alkaloids.

In general, most of the angiosperm plants are susceptible to *Agrobacterium*-mediated transformation and production of hairy roots. Successful genetic transformation and establishment of hairy root cultures has been reported for several genera of TA-producing families. In Solanaceae, different genera including *Atropa*,

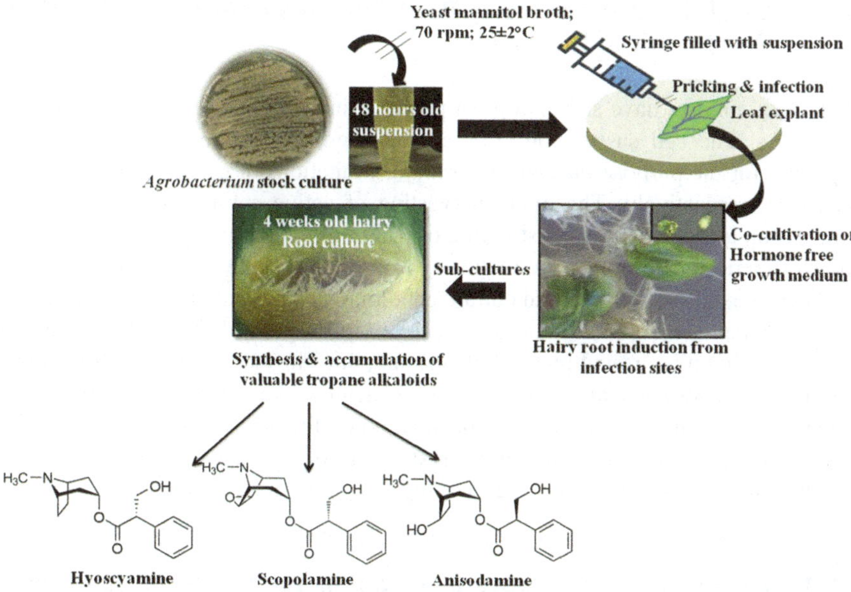

Fig. 5.1 General protocol for hairy root induction in tropane alkaloid-bearing plants and production of valuable tropane alkaloids

Anisodus, *Brugmansia*, *Datura*, *Duboisia*, *Hyoscyamus*, and *Scopolia* are known to produce hairy roots upon infection with different strains of *Agrobacterium*. A relatively simple method of direct infection and cocultivation of sterile explants with *A. rhizogenes* appears to be the most effective technique for hairy root induction. Simple tissue culture laboratory conditions with no specific equipment or laboratory setup are required for this experimentation to induce hairy roots (Fig. 5.1). Thoroughly sterilized explants (any part of the plant; very often leaves) are superficially wounded with syringe or scalpel and inoculated with *A. rhizogenes* suspension, normally 2 days' old. Several *Agrobacterium* strains including *A. rhizogenes* ATCC 15834, ATCC13333, 8196, 1724, 2659, 1855, AR-10, TR-105, A41027, A4, LBA 9402, LBA 1334, R1601, and MAFF 03–01724 have been reported to induce hairy roots in different plant parts of various members of TA-producing plant families. Besides, *Agrobacterium tumefaciens* C58 C1 carrying the pRi 15834 plasmid and double transformation with two *A. rhizogenes* strains (ATCC 15834 and MAFF 03–01724) are also known to produce hairy roots in Solanaceae members (Bonhomme et al. 2000; Akramian et al. 2008; Shakeran et al. 2014).

5.3 Establishment of Hairy Root Cultures of Some TA-Bearing Plants: Initial Reports

Hairy root cultures have shown a great potential in biotechnological production of plant TAs. The cell suspension cultures and chemical synthesis have also been exploited for the purpose but only with major limitations of genetic instability and high cost, respectively. The direct harvesting of active compounds from natural sources can lead to complete destruction of plants. On the contrary, hairy roots have shown superiority over other production systems in terms of easy establishment and maintenance, hormone-free basal culture conditions, genetic and biochemical stability, and potential to upscale in a large-scale setup. In general, TAs synthesize and accumulate in underground parts of plants, and, thus, in such cases, in vitro exploitation of root systems appears to be more genuine and beneficial. Almost all members of Solanaceae family have been explored for their hairy root cultures for native TA production. In consecutive text, some of the randomly selected plant members have been described for exploiting their hairy root cultures.

5.3.1 Atropa, Anisodus, Datura, Scopolia, and Duboisia Hairy Root Cultures as Examples

Atropa is considered as classic and most cited example of TA-producing plant of which hairy root cultures have been generated and used for the production of not only important metabolites but also various other hairy root-based applications. These include biotransformation of active compounds into their more useful chemical form and remediation of toxic compounds utilizing root bore enzymatic system (Srivastava et al. 2012; Mazaheri and Piri 2015; Habibi et al. 2018). *Atropa* hairy roots have also been used to screen transformants for the expression of foreign genes (Banerjee et al. 2002). Initially, hairy root culture establishment protocol for *A. belladonna* was provided by Bonhomme et al. (2000) by cocultivation with two strains of *A. tumefaciens* harboring *A. rhizogenes rol*C gene alone or *rol*ABC genes together (Bonhomme et al. 2000). Several hairy root lines have been obtained from the infection of two bacterial strains with higher hyoscyamine and scopolamine levels (0.85% DW to 0.4% DW) when compared to normal and untransformed roots. Other *A. rhizogenes* strains like AR15834, A4, and LBA9402 have also been used to initiate hairy root cultures from *Atropa* (Aoki et al. 1997 and reference therein]. Besides *A. belladonna*, other species of *Atropa*, viz., *A. accuminata*, has also been reported for exploitation of its hairy root cultures and tropane alkaloid profiling (Banerjee et al. 2008).

Anisodus acutangulus, an endemic plant to China, is an important TA-producing medicinal plant of which roots have been used for medicinal purposes since long. Hairy roots of this plant have been worked out for TA production. Kai et al. reported the induction of hairy roots from different parts of in vitro-grown seedlings upon infection with disarmed *A. tumefaciens* strain C58C1 possessing two plasmids, *A. rhizogenes* Ri plasmid pRiA4 and another plasmid containing a target gene

(Li et al. 2008a, b; Kai et al. 2018). A brief coculture period followed by gradual treatment of antibiotic cefotaxime for 14 days resulted in several fast-growing hairy root clones. These clones after confirmation were investigated for their TA production potential.

Datura, another important genus from the Solanaceae family, has been widely reported and exploited for its TAs, particularly hyoscyamine. Three *Datura* species (*D. stramonium*, *D. tatula*, and *D. innoxia*) have been investigated for their hairy root cultures (Harfi and Khelifi 2019). The hairy root induction in these species was done through the infection and cocultivation A4 strain of *A. rhizogenes* with an agropine-type plasmid. However, initial studies on hairy root cultures of *Datura stramonium* include a report by Maldonado-Mendoza et al. (1993) about establishment and characterization of hairy roots of this species. The study also included the factors responsible for the biochemical stability of these hairy roots during prolonged period of subculturing (Maldonado-Mendoza et al. 1993).

Scopolia japonica also known as Japanese belladonna is another medicinally important member of nightshade family. Besides TAs, the plant roots are also known to produce coumarins, umbelliferone, and scopoletin. In *Scopolia* genus, initial studies of hairy root induction and establishment were reported by Mano et al. (1986). Twenty-nine hairy root clones of *Scopolia japonica* were established through the infection and coculture of *A. rhizogenes* strain 15,834. All isolated hairy root lines exhibited various phenotypic and biochemical traits. Of these, two high-yielding alkaloid productive clones were investigated for their growth rate and tropane alkaloid production potential under varying cultural conditions.

Duboisia leichhardtii and *D. myoporoides* of the family Solanaceae are indigenous plants to Australia and New Caledonia. The plants contain tropane alkaloids and also were reported to have significant amount of pyridine alkaloids. Scopolamine and hyoscyamine are the two main TAs present in the genus *Duboisia* which provide the plant a high medicinal value. Among the initial reports, *D. leichhardtii* was reported to produce hairy roots from leaf disks and stem segments infected with *A. rhizogenes* strains 15834 and A4 (Mano et al. 1989; Muranaka et al. 1993; Yoshimatsu et al. 2004).

Leaf explants of hybrid of *D. myoporoides* and *D. leichhardtii* are also reported for inducing hairy roots through the infection and coculture of ATCC 15834 (Yoshimatsu et al. 2004). The study was aimed to investigate the effect of bacterial density and coculture conditions on hybrid hairy root induction. Later, hairy roots were chemically analyzed and reported to accumulate much higher TAs (35 mg/l scopolamine and 17 mg/l hyoscyamine) when compared to adventitious roots grown under similar culture medium and other conditions except phytohormone supplementation (0.5 mg/l IAA).

5.4 Biotechnological Strategies Used to Maximize the Production of Tropane Alkaloids from Hairy Roots

Although hairy root cultures have shown their worth for biotechnological production of plant TAs yet, there is always the limitation of low productivity when compared to the large experimental setups. Further, in light of heavy pharmaceutical and other commercial demand, there is a need to incorporate certain productivity enhancement measures utilizing suitable hairy root lines. In hairy root-based production of plant TAs, several biotechnological strategies have been employed to enhance the productivity of root system. The coming text discusses these strategies with relevant examples.

5.4.1 Preliminary Strategies

5.4.1.1 Selection of Fast-Growing, High-Yielding Hairy Root Line, Media Optimization, and Assessment of Effect of Bacterial Strain

Emergence of hairy root is a direct consequence of genetic transformation of host plant cell by *Agrobacterium* plasmid (Chandra 2012). The part of plants to which transformed cell belongs; age and physiological state of host cell are among some of the host factors which directly influence the event of transformation and emergence of root. Further, influence of tDNA integration event into host cell genome (in terms of integration site and number of tDNA copies inserted and their orientation as direct or inverted repeat) is also among the decisive factors that influence the morphological characteristics and biochemical potential of different hairy root lines (Mehrotra et al. 2020 and reference therein). Moreover, in plant cell cultures also, due to the possibilities of somaclonal variations, there has been a common practice of high secondary metabolite-yielding cell line selection (Krishna et al. 2016). On this basis, as and when hairy roots of a test plant system appears on infection sites of explant, they are considered and cultured as individual lines. After a thorough biochemical analysis, high-yielding hairy root lines are selected for further scientific endeavors. Other studies which substantiate this fact include establishment and selection of high tropane alkaloid-yielding hairy root lines of *Datura candida*, *D. stramonium*, and somatic hybrid of *Hyoscyamus albus* and *H. muticus* (Christen et al. 1989; Zehra et al. 1998; Harfi et al. 2018). In *Atropa* hairy roots, several clones have been selected and characterized as high alkaloid-yielding root clones (Eskandari-Samet et al. 2012).

Unlike cell suspension cultures, biochemical potential of hairy root clones does not get affected with physical and chemical conditions of growth. However, several studies have been performed to assess the effect of medium constituents on growth and metabolite production of hairy roots of various TA-producing plants (Oksman-Caldentey et al. 1994; Liu et al. 2013). Tropane alkaloid production of *Anisodus acutangulus* and *H. niger* hairy roots has been investigated through the optimization of carbon source, nitrogen, and other inorganic ion concentration and culture medium pH (Li et al. 2012; Pudersell et al. 2012). Besides, hairy roots of

A. belladonna and *D. stramonium* are also subjected to the analysis of medium constituents and tropane alkaloid production profile (Pinol et al. 1999; Pudersell et al. 2007).

Selection of *A. rhizogenes* strain for effective transformation, hairy root induction, and alkaloid productivity is also crucial. In *H. albus* hairy roots produced by *Agrobacterium* strain A4, the atropine yields are much higher than that of roots produced by the infection of strain LBA9402 (Zehra et al. 1999). However, in similar exercise with *H. muticus* infected with bacterial strain A4 and LBA9402, no significant difference in alkaloid yield was observed. Such studies suggest the influence of specific bacterial strain-host plant specificity in tropane alkaloid production through hairy roots, even in the members of the same genera of one family.

5.4.2 Additional Strategies

5.4.2.1 Elicitation, Genetic/Metabolic Engineering, and Bioreactor Scale-Up

Elicitation is known as a practically most feasible strategy for the enhancement of secondary metabolite production from cell and tissue cultures (Ramirez-Estrada et al. 2016). Generally, any biotic or abiotic compound or physical condition which triggers the synthesis or accumulation of secondary metabolites in source tissue is considered as elicitor, and the process is known as elicitation (Zhao et al. 2005; Goel et al. 2011). Elicitor molecules are capable to induce phytoalexin synthesis and trigger defense response in host plants primarily in terms of secondary metabolite synthesis (Halder et al. 2019). In previous decades, elicitation has been used as an effective method to stimulate secondary metabolite production or their efflux (in growth medium) in differentiated and/or undifferentiated in vitro cultures of several plant species (Poornananda and Al-Khayri 2016). Biotic and abiotic elicitation has also been a common strategy for metabolite production and/or their yield enhancement in hairy root cultures (Halder et al. 2019 and reference therein).

A range of biotic and abiotic elicitors have been employed to increase the productivity of hairy root cultures of TA-producing plants. Various biotic and abiotic elicitors like salicylic acid, $AgNo_3$, $CaCl_2$, and $CdCl_2$ have been used in hairy roots of *Brugmansia candida* for the enhancement of scopolamine and hyoscyamine (Pitta et al. 2000; Spollansky et al. 2000). In hairy root cultures of *H. reticulatus*, iron oxide nanoparticles were used for elicitation which resulted in enhanced accumulation of scopolamine and hyoscyamine (Moharrami et al. 2017). Similarly, Ag^+, ethanol, and MJ have been used to elevate the production of native TAs in *Anisodus acutangulus* (Kai et al. 2012a, b). Extensive studies of elicitation in hairy root cultures of different *Datura* species have been performed. MJ, SA, and acetyl salicylic acid have been used as elicitors for hairy root cultures of *D. stramonium* and *Datura tatula* hairy root lines for the TA production particularly scopolamine and hyoscyamine (Sun et al. 2013; Harfi et al. 2018). 30-day and 60-day treatment of MJ induced the high production of scopolamine and hyoscyamine in *D. stramonium* hairy root that have leached out into the culture medium.

However, the results in *D. tatula* revealed significant high amount (17.94 mg/g DW) of hyoscyamine in hairy roots treated with 0.1 mM acetyl salicylic acid. KCL and $CaCl_2$ as chemical elicitors have been used to investigate accumulation of TAs in hairy root lines of *Datura stramonium*, *D. innoxia*, and *D. tatula* (Harfi et al. 2016). The concentrations and exposure time of the elicitor had a considerable effect on hyoscyamine accumulation in these hairy roots. In a recent study of *Hyoscyamus reticulatus* L. hairy root cultures, sodium nitroprusside (SNP), a nitric oxide donor, was used as elicitor in various concentrations (0 μM, 50 μM, and 300 μM) and exposure times (24 and 48 h) (Khezerluo et al. 2018). The result exhibited highest hyoscyamine (1.2-fold) and scopolamine (1.5-fold) production at 50 and 100 μM SNP at 48 and 24 h of exposure time, respectively.

Biotic elicitors have also been frequently used in hairy roots of TA-producing plants. Supplementation of *Bacillus cereus* and *Staphylococcus aureus* extract as biotic elicitors was used and investigated for the production of scopolamine in *D. metel* hairy roots (Shakeran et al. 2017). Result showed improvement in scopolamine production but reduction in atropine content. Elicitation by *B. cereus* caused production of scopolamine (0.03 g and 0.017 g/100 g DW of hairy root) after 12 and 24 h respectively. However, hairy roots elicited with *S. aureus* did not produced scopolamine after 12 h, but 0.025 g/100 g DW scopolamine was detected after 24 h. Earlier to this, in an attempt of productivity enhancement of hairy root cultures of *Scopolia parviflora*, two Gram-positive and one Gram-negative strain of bacteria were used as biotic elicitors (Jung et al. 2003). The scopolamine content was increased with raw bacterial elicitors, whereas the autoclaved bacterial elicitors exhibited similar effects.

Genetic/metabolic engineering has gained popularity as an investigative as well as productivity enhancement strategy for in vitro cultures. As the information about the biosynthetic pathways of several bioactive compounds and genes involved within become clear, the efforts for engineering pathway steps have gained scientific focus. In TA-producing plants, as the biosynthetic pathway for TAs is nearly known, a number of pathway genes of rate limiting have been engineered to modulate the product flux (Kai et al. 2018). Several TA biosynthetic pathway genes have been isolated, cloned, and characterized in different TA-producing plant species including *A. belladonna*, *Hyoscyamus* spp., *Datura* spp., and *Anisodus acutangulus*.

Single or multistep genetic engineering strategy has been employed for the enhancement of hairy root-based TA production. In early studies of *Duboisia* plants, CaMV 35S-controlled tobacco PMT gene was overexpressed (Moyano et al. 2002). PMT (putrescine N-methyltransferase) is the member of S-adenosyl-methionine-dependent N-methylation transferase family and responsible for the first rate-limiting step of TA biosynthetic pathway. Later, in hairy roots of *H. niger*, overexpression of PMT was also carried out (Zhang et al. 2007). However, in this study, simple overexpression of PMT could only lead to the increase in methylputrescine (MPUT, an intermediate compound) but not the TAs. However, treatment of MeJA in these hairy roots has led to a significant increase in scopolamine content (Zhang et al. 2007). In *Atropa baetica* and *A. belladonna*, overexpression of *h6h* that codes for enzyme hyoscyamine6-β-hydroxylase (H6H) has resulted in increased content of scopolamine (Zarate et al. 2006). In *H. muticus* L. strain Cairo, overexpression of *35S-h6h* has resulted in increased production of scopolamine (Jouhikainen et al. 1999). Tropine reductases (TRI and TRII) that have been characterized from various TA-bearing plants have also been investigated for their role in *Scopolia lurida* and *Anisodus acutangulus* hairy root cultures (Zhao et al. 2017). The overexpression of SlTRI and AaTRI in these hairy roots has led to the accumulation of hyoscyamine of 1.7–2.0 and 1.87-fold high, respectively in comparison to control (Kai et al. 2018).

Similarly, in some studies, multistep genetic engineering approach has been employed for TA yield enhancement in hairy root cultures. In two simultaneous studies, Kai et al. (2011, 2012a, b) investigated overexpression of AaPMT + AaTRI and AaTRI + H6H in hairy roots of *Anisodus acutangulus* (Kai et al. 2011, 2012a, b). PMT is considered as the first upstream rate-limiting enzyme, whereas TRI and H6H are branch controlling and downstream rate-limiting enzymes in tropane alkaloid biosynthetic pathway. In similar kind of studies, overexpression of PMT and H6H in *A. belladonna*, *Scopolia parviflora*, and *H. muticus* hairy roots has shown considerable improvement in hyoscyamine and scopolamine content (Zhang et al. 2004; Kang et al. 2011; Yang et al. 2011). These studies have concluded in increased production of native TAs through overexpression of single or more than one pathway genes and have suggested genetic engineering as a potential strategy for TA production in hairy roots.

Bioreactor scale-up of secondary metabolite-producing hairy root cultures has also attracted wide scientific attention. Various studies have been performed on reactor designing and configuration for upscaling of hairy roots (Mehrotra et al. 2016). How to upscale a production system to harness its commercial benefits is always been a matter of research and practice. Considering hairy root culture systems as a feasible production platform, the efforts are being made to develop a suitable large-scale culture production system for the desired and commercially required plant metabolites (Gerth et al. 2007; Mishra and Ranjan 2008; Stiles and Liu 2013). In an earlier study, upscale culture of *D. stramonium* hairy roots in a modified stirred tank reactor exhibited hyoscyamine levels approximately 0.5 mg/g FW (Hilton and Rhodes 1990). Upscaling of hairy roots of *H. niger* in a traditional, modified bubble column (spray) bioreactor has been reported. In a comparison, anisodine content (0.67 mg/g DW) was obtained higher in modified reactor than scopolamine and hyoscyamine (5.3 mg/g and 1.6 mg/g DW, respectively) which were found higher in traditional bubble column reactor (Jaremicz et al. 2014). Results of the study of upscaling of *A. belladonna* hairy roots in a 1.5 l bioreactor using the inoculum size of 0.5 gFW revealed maximum scopolamine production (1.59 mg/g^{-1} DW) (Habibi et al. 2015).

5.5 Conclusion

In conclusion, the huge contribution of plant-derived tropane alkaloids having therapeutic importance in the field of pharmaceuticals for making novel drugs has resulted in the additional attention of researchers toward isolation, extraction, and enhancement of the production of bioactive compounds by using different biotechnological interventions. Biotechnology offers many choices by which secondary metabolism in medicinal plants can be distorted in many ways. Hairy root culture is one of the most effective and precious tool to fulfill the increasing commercial demand of the tropane alkaloids. As the TAs are the oldest known compounds for medicinal use, their production through novel tissue culture technique by means of *Agrobacterium*-mediated gene transfer is very demanding and challenging. Owing to this, the present chapter has focused on describing the preliminary and additional strategies to enhance the productivity of hairy root system by using various biotechnological approaches like selection of fast-growing, high-yielding hairy root line, media optimization and assessment of the effect of bacterial strain, elicitation, genetic/metabolic engineering, bioreactor scale-up, etc.

References

Ahmad S, Garg M, Tamboli ET, Abdin MZ, Ansari SH (2013) In vitro production of alkaloids: factors, approaches, challenges and prospects. Pharmacogn Rev 7(13):27–33. https://doi.org/10.4103/0973-7847.112837

Akramian M, Tabatabaei SMF, Mirmasoumi M (2008) Virulence of different strains of *Agrobacterium rhizogenes* on genetic transformation of four Hyoscyamus species. Amer-Euras. J Agric Environ Sci 3(5):759–763

Aoki T, Matsumoto H, Asako Y, Matsunaga Y, Shimomura K (1997) Variation of alkaloid productivity among several clones of hairy roots and regenerated plants of *Atropa belladonna* transformed with *Agrobacterium rhizogenes* 15834. Plant Cell Rep 16(5):282–286. https://doi.org/10.1007/BF01088281

Banerjee S, Shang TQ, Wilson AM, Moore AL, Strand SE, Gordon MP, Doty SL (2002) Expression of functional mammalian P450 2E1 in hairy root cultures. Biotechnol Bioeng 77 (4):462–466. https://doi.org/10.1002/bit.10151

Banerjee S, Madhusudanan KP, Chattopadhyay SK, Rahman LU, Khanuja SPS (2008) Expression of tropane alkaloids in the hairy root culture of *Atropaacuminata* substantiated by DART mass spectrometric technique. Biomed Chromatogr 22(8):830–834. https://doi.org/10.1002/bmc.998

Bonhomme V, Laurain-Mattar D, Lacoux J, Fliniaux M, Jacquin-Dubreuil A (2000) Tropane alkaloid production by hairy roots of *Atropa belladonna* obtained after transformation with *Agrobacterium rhizogenes* 15834 and *Agrobacterium tumefaciens* containing rol A, B, C genes only. J Biotechnol 81:151–158. https://doi.org/10.1016/s0168-1656(00)00287-x

Chandra S (2012) Natural plant genetic engineer *Agrobacterium rhizogenes*: role of T-DNA in plant secondary metabolism. Biotechnol Lett 34:407–415. https://doi.org/10.1007/s10529-011-0785-3

Chilton MD, Tepfer DA, Petit A, David CY, Casse Delbart F, Tempe J (1982) *Agrobacterium rhizogenes* inserts T-DNA into the genomes of the host plant root cells. Nature 295 (5848):432–434. https://doi.org/10.1038/295432a0

Christen P, Roberts MF, Phillipson JD, Evans WC (1989) High-yield production of tropane alkaloids by hairy-root cultures of a *Datura candida* hybrid. Plant Cell Rep 8:75–77. https://doi.org/10.1007/BF00716842

Eskandari-Samet A, Piri K, Kayhanfar M, Hasanloo T (2012) Enhancement of tropane alkaloid production among several clones and explants types of hairy root of *Atropa belladonna* L. J Med Plant By-Prod 1:35–42

Gelvin SB (2009) Agrobacterium in the genomics age. Plant Physiol 150:1665–1676. https://doi.org/10.1104/pp.109.139873

Georgiev MI, Agostini E, Ludwig-Muller J, Xu J (2012) Genetically transformed roots: from plant disease to biotechnological resource. Trends Biotechnol 30(10):528–537. https://doi.org/10.1016/j.tibtech.2012.07.001

Georgiev V, Marchev A, Berkov S, Pavlov A (2013) Plant in vitro systems as sources of tropane alkaloids. In: Ramawat K, Merillon JM (eds) Natural products. Springer, Berlin, Heidelberg. https://doi.org/10.1007/978-3-642-22144-6_6

Gerth A, Schmidt D, Wilken D (2007) The production of plant secondary metabolites using bioreactors. Acta Hortic 764:95–104. https://doi.org/10.17660/ActaHortic.2007.764.11

Goel MK, Mehrotra S, Kukreja AK (2011) Elicitor-induced cellular and molecular events are responsible for productivity enhancement in hairy root cultures: an insight study. Appl Biochem Biotechnol 165(5–6):1342–1355. https://doi.org/10.1007/s12010-011-9351-7

Gonçalves S, Anabela R (2018) Production of plant secondary metabolites by using biotechnological tools. In: Ramasamy V, SSS R (eds) Secondary metabolites—sources and applications. IntechOpen, London. https://doi.org/10.5772/intechopen.76414

Goyal S (2013) Ecological role of alkaloids. In: Ramawat K, Merillon JM (eds) Natural products. Springer, Berlin, Heidelberg

Gutierrez-Valdes N, Hakkinen ST, Lemasson C, Guillet M, Oksman-Caldentey KM, Ritala A, Cardon F (2020) Hairy root cultures-a versatile tool with multiple applications. Front Plant Sci 11:33. https://doi.org/10.3389/fpls.2020.00033

Habibi P, Piri K, Ali D, Moghadam YA, Ghiasvand T (2015) Increasing scopolamine content in hairy roots of *Atropa belladonna* using bioreactor. Braz Arch Biol Technol 58(2):166–174. https://doi.org/10.1590/S1516-8913201400276

Habibi P, Soccol CR, Grossi-de-Sa MF (2018) Hairy root-mediated biotransformation: recent advances and exciting prospects. In: Srivastava V et al (eds) Hairy roots: an effective tool of plant biology. Springer, Singapore, pp 71–93

Halder M, Sarkar S, Jha S (2019) Elicitation: a biotechnological tool for enhanced production of secondary metabolites in hairy root cultures. Eng Life Sci 19:880–895. https://doi.org/10.1002/elsc.201900058

Harfi B, Khelifi L (2019) Hairy root technology: effect of etiolation on Datura sp. transgenic root induction and Hyoscyamine production. Pharm J 11(5):991–995. https://doi.org/10.5530/pj.2019.11.156

Harfi B, Khelifi-Slaoui M, Bekhouche M, Benyammi R, Hefferon K, Makhzoum A, Khelifi L (2016) Hyoscyamine production in hairy roots of three Datura species exposed to high-salt medium. In Vitro Cell Dev Biol – Plant 52:92–98. https://doi.org/10.1007/s11627-015-9725-6

Harfi B, Khelifi L, Khelifi-Slaoui M, Assaf-Ducrocq C, Gontier E (2018) Tropane alkaloids GC/MS analysis and low dose elicitors' effects on hyoscyamine biosynthetic pathway in hairy roots of Algerian Datura species. Sci Rep 8:17951. https://doi.org/10.1038/s41598-018-36625-4

Hasan, Hasibul Q (2017) Tropane—molecule of the month June 2012 [archived version]. Figshare J. https://doi.org/10.6084/m9.figshare.5414788.v1

Hilton MG, Rhodes MJC (1990) Growth and hyoscyamine production by hairy root cultures of *Datura stramonium* in a modified stirred tank reactor. Appl Microbiol Biotechnol 33:132–138. https://doi.org/10.1007/BF00176513

Jaremicz Z, Luczkiewicz M, Kokotkiewicz A, Krolicka A, Sowinski P (2014) Production of tropane alkaloids in *Hyoscyamus niger* (black henbane) hairy roots grown in bubble-column and spray bioreactors. Biotechnol Lett 36(4):843–853. https://doi.org/10.1007/s10529-013-1426-9

Jirschitzka J, Schmidt GW, Reichelt M, Schneider B, Gershenzon J, D'Auria JC (2012) Plant tropane alkaloid biosynthesis evolved independently in the Solanaceae and Erythroxylaceae. Proc Natl Acad Sci 109(26):10304–10309. https://doi.org/10.1073/pnas.1200473109

Jouhikainen K, Lindgren L, Jokelainen T, Hiltunen R, Teeri TH, Oksman-Caldentey KM (1999) Enhancement of scopolamine production in *Hyoscyamus muticus* L. hairy root cultures by genetic engineering. Planta 208(4):545–551

Jung HY, Kang SM, Kang YM, Kang MJ, Yun DJ, Bahk JD, Yang JK, Choi MS (2003) Enhanced production of scopolamine by bacterial elicitors in adventitious hairy root cultures of *Scopoliaparviflora*. Enzyme Microb Technol 33:987–990

Kai G, Yang S, Luo X, Zhou W, Fu X, Zhang A, Zhang Y, Xiao J (2011) Co-expression of AaPMT and AaTRI effectively enhances the yields of tropane alkaloids in *Anisodusacutangulus* hairy roots. BMC Biotechnol 11:43. https://doi.org/10.1186/1472-6750-11-43

Kai G, Zhang A, Guo Y, Li L, Cui L, Luo X, Xiao J (2012a) Enhancing the production of tropane alkaloids in transgenic *Anisodusacutangulus* hairy root cultures by over-expressing tropinonereductase I and hyoscyamine-6β-hydroxylase. Mol Bio Syst 8(11):2883–2890. https://doi.org/10.1039/c2mb25208b

Kai G, Yang S, Zhang Y, Luo X, Fu X, Zhang A, Xiao J (2012b) Effects of different elicitors on yield of tropane alkaloids in hairy roots of *Anisodusacutangulus*. Mol Biol Rep 39 (2):1721–1729. https://doi.org/10.1007/s11033-011-0912-1

Kai G, Zhao W, Shi M, Wang Y (2018) Biotechnological interventions of hairy roots of tropane alkaloid bearing plants. In: Srivastava V et al (eds) Hairy roots: an effective tool of plant biology. Springer, Singapore, pp 71–93

Kang YM, Park DJ, Min JY, Song HJ, Jeong MJ, Kim YD, Choi MS (2011) Enhanced production of tropane alkaloids in transgenic *Scopoliaparviflora* hairy root cultures over-expressing putrescine N-methyl transferase (PMT) and hyoscyamine-6β-hydroxylase (H6H). In Vitro Cell Dev-Pl 47(4):516–524. https://doi.org/10.1007/s11627-011-9367-2

Khezerluo M, Hosseini B, Amiri J (2018) Sodium nitroprusside stimulated production of tropane alkaloids and antioxidant enzymes activity in hairy root culture of *Hyoscyamus reticulatus* L. Acta Biol Hung 69(4):437–448. https://doi.org/10.1556/018.69.2018.4.6

Kim N, Estrada O, Chavez B, Stewart C Jr, D'Auria JC (2016) Tropane and Granatane alkaloid biosynthesis: a systematic analysis. Molecules 21(11):1510. https://doi.org/10.3390/molecules21111510

Kohnen-Johannsen KL, Kayser O (2019) Tropane alkaloids: chemistry, pharmacology, biosynthesis and production. Molecules 24(4):796. https://doi.org/10.3390/molecules24040796

Krishna H, Alizadeh M, Singh D, Singh U, Chauhan N, Eftekhari M, Sadh R (2016) Somaclonal variations and their applications in horticultural crops improvement. 3 Biotech 6(1):54. https://doi.org/10.1007/s13205-016-0389-7

Li L, Wang J, Wang W, Lu Y, Wang YL, Zhou GY, Kai GY (2008a) Optimization of induction and culture conditions and tropane alkaloid production in hairy roots of *Anisodusacutangulus*. Biotechnol Bioprocess Eng 13:606–612

Li L, Wang J, Wang W, LuY WYL, Zhou GY, Kai GY (2008b) Optimization of induction and culture conditions and tropane alkaloid production in hairy roots of *Anisodusacutangulus*. Biotechnol Bioprocess Eng 13:606–612

Liu Q, Cui L, Guo Y, Ni X, Zhang Y, Kai G (2013) Optimization of nutritive factors in culture Media for Growth and Tropane Alkaloid Production from Anisodusacutangulus hairy roots. J App Pharmaceut Sci 3(1):001–004. https://doi.org/10.7324/JAPS.2013.30101

Maldonado-Mendoza IE, Ayora-Talavera T, Loyola-Vargas VM (1993) Establishment of hairy root cultures of *Datura stramonium*. Characterization and stability of tropane alkaloid production during long periods of subculturing. Plant Cell Tiss Org Cult 33:321–329. https://doi.org/10.1007/BF02319018

Mano Y, Nabeshima S, Chiaki M, Ohkawa H (1986) Production of tropane alkaloids by hairy root cultures of *Scopolia japonica*. Agric Biol Chem 50(1):2715–2722. https://doi.org/10.1271/bbb1961.50.2715

Mano Y, Ohkawa H, Yamada Y (1989) Production of tropane alkaloids by hairy root cultures of *Duboisia leichhardtii* transformed by *Agrobacterium rhizogenes*. Plant Sci 59 (2):191–201. https://doi.org/10.1016/0168-9452(89)90137-4

Mazaheri H, Piri K (2015) Removal of phenol by *A. belladonna* L. hairy root. Int J Phytoremediation 17(12):1212–1219. https://doi.org/10.1080/15226514.2015.1045136

Mehrotra S, Srivastava V, Rahman LU, Kukreja AK (2015) Hairy root biotechnology—indicative timeline to understand missing links and future outlook. Protoplasma 252:1189–1201. https://doi.org/10.1007/s00709-015-0761-1

Mehrotra S, Mishra S, Srivastava V (2016) Bioreactor technology for hairy roots cultivation. In: Pavlov A, Bley T (eds) Bioprocessing of plant in vitro systems, reference series in phytochemistry. Springer, Singapore. https://doi.org/10.1007/978-3-319-32004-5_10-1

Mehrotra S, Mishra S, Srivastava V (2020) Hairy roots biotechnology unzipped: a journey of reality and promises. In: Srivastava V et al (eds) Hairy root cultures based applications: methods and protocols. Springer, Singapore. https://www.springer.com/gp/book/9789811540547

Mishra BN, Ranjan R (2008) Growth of hairy root cultures in various bioreactors for the production of secondary metabolites. Biotechnol Appl Biochem 49(1):1–10. https://doi.org/10.1042/BA20070103

Moharrami F, Hosseini B, Sharafi A, Farjaminezhad M (2017) Enhanced production of hyoscyamine and scopolamine from genetically transformed root culture of *Hyoscyamus reticulatus* L. elicited by iron oxide nanoparticles. In Vitro Cell Dev Biol Plant 53(2):104–111. https://doi.org/10.1007/s11627-017-9802-0

Moyano E, Fornale S, Palazon J, Cusido RM, Bagni PMT (2002) Alkaloid production in Duboisia hybrid hairy root cultures overexpressing the pmt gene. Phytochemistry 59(7):697–702. https://doi.org/10.1016/s0031-9422(02)00044-4

Muranaka T, Ohkawa H, Yamada Y (1993) Continuous production of scopolamine by a culture of *Duboisia leichhardtii* hairy root clone in a bioreactor system. Appl Microbiol Biotechnol 40:219–223. https://doi.org/10.1007/BF00170370

Oksman-Caldentey KM, Arroo R (2000) Regulation of tropane alkaloid metabolism in plants and plant cell cultures. In: Verpoorte R, Alfermann AW (eds) Metabolic engineering of plant secondary metabolism. Springer, Dordrecht

Oksman-Caldentey K, Sevon N, Vanhala L, Hiltunen R (1994) Effect of nitrogen and sucrose on the primary and secondary metabolism of transformed root cultures of *Hyoscyamus muticus*. Plant Cell Tiss Org Cult 38:263–272. https://doi.org/10.1007/BF00033886

Paul R, Datta KA (2011) An updated overview on Atropa belladonna L. Int Res J Pharmacy 2 (11):11–17

Pinol MT, Palazon J, Cusido RM, Ribo M (1999) Influence of calcium ion concentration in the medium on tropane alkaloid accumulation in *Datura stramonium* hairy roots. Plant Sci 141:41–49

Pitta S, Spollansky T, Ana G (2000) The influence of different biotic and abiotic elicitors on the production and profile of tropane alkaloids in hairy root cultures of *Brugmansia candida*. Enzyme Microb Technol 26:252–258. https://doi.org/10.1016/S0141-0229(99)00137-4

Poornananda MN Jameel Al-Khayri (2016) Abiotic and biotic elicitors–role in secondary metabolites production through in vitro culture of medicinal plants. Abiotic and biotic stress in plants—recent advances and future perspectives, Arun K. Shanker and Chitra Shanker, Intech Open, London doi:https://doi.org/10.5772/61442

Pudersell K, Vardja R, Vardja T, Arak E, Raal A (2007) Influence of calcium on the tropane alkaloid production and riboflavine excretion in the root cultures of Hyoscyamus niger L. Eur J Pharm Sci 32:S33–S34

Pudersell K, Vardja T, Vardja R, Matto V, Arak E, Raal A (2012) Inorganic ions in the medium modify tropane alkaloids and riboflavin output in *Hyoscyamus niger* root cultures. Pharmacogn Mag 8(29):73–77. https://doi.org/10.4103/0973-1296.93330

Ramirez-Estrada K, Vidal-Limon H, Hidalgo D, Moyano E, Golenioswki M, Cusido RM, Palazon J (2016) Elicitation, an effective strategy for the biotechnological production of bioactive high-added value compounds in plant cell factories. Molecules 21(2):182. https://doi.org/10.3390/molecules21020182

Shakeran Z, Keyhanfar M, Gholamreza A (2014) Hairy roots formation in four Solanaceae species by different strains of *Agrobacterium rhizogenes*. J Med Plants By-products 2:155–160

Shakeran Z, Keyhanfar M, Ghanadian M (2017) Biotic elicitation for scopolamine production by hairy root cultures of *Daturametel*. Mol Biol Res Commun 6(4):169–179. https://doi.org/10.22099/mbrc.2017.25776.1275

Spollansky TC, Pitta-Alvarez SI, Giulietti AM (2000) Effect of jasmonic acid and aluminium on production of tropane alkaloids in hairy root cultures of *Brugmansia candida*. Electron J Biotechnol 3(1):31–32. https://doi.org/10.2225/vol3-issue1-fulltext-6

Srivastava V, Negi AS, Ajayakumar PV, Khan AS, Banerjee S (2012) *Atropa belladonna* hairy roots: orchestration of concurrent oxidation and reduction reactions for biotransformation of carbonyl compounds. Appl Biochem Biotechnol 166(6):1401–1408. https://doi.org/10.1007/s12010-011-9533-3

Stiles AR, Liu CZ (2013) Hairy root culture: bioreactor design and process intensification. Adv Biochem Eng Biotechnol 134:91–114. https://doi.org/10.1007/10-2013-181

Sun JW, Zhang H, Wang FY, Sun YM, Sun M (2013) Effects of methyl jasmonate on accumulation and release of main tropane alkaloids in liquid cultures of *Datura stramonium* hairy root. China J Chin Mater Medica 38(11):1712–1718

Yang CX, Chen M, Zeng LJ, Zhang L, Liu XQ, Lan XZ, Tang KX, Liao Z (2011) Improvement of tropane alkaloids production in hairy root cultures of *Atropa belladonna* by overexpressing pmt and h6h genes. Plant Omics J 4(1):29–33

Yoshimatsu K, Sudo H, Kamada H, Kiuchi F, Kikuchi Y, Sawada JI, Shimomura K (2004) Tropane alkaloid production and shoot regeneration in hairy and adventitious root cultures of *Duboisia myoporoides-D. leichhardtii* hybrid. Biol Pharm Bull 27(8):1261–1265. https://doi.org/10.1248/bpb.27.1261

Zarate R, el Jaber-Vazdekis N, Medina B, Ravelo AG (2006) Tailoring tropane alkaloid accumulation in transgenic hairy roots of *Atropabaetica* by over-expressing the gene encoding hyoscyamine 6β-hydroxylase. Biotechnol Lett 28(16):1271–1277. https://doi.org/10.1007/s10529-006-9085-8

Zehra M, Banerjee S, Naqvi AA, Kumar S (1998) Variation in the growth and alkaloid production capability of the hairy roots of *Hyoscyamus albus, H. muticus* and their somatic hybrid. Plant Sci 136(1):93–99

Zehra M, Banerjee S, Sharma S, Kumar S (1999) Influence of *Agrobacterium rhizogenes* strains on biomass and alkaloid productivity in hairy root lines of *Hyoscyamus muticus* and *H albus*. Planta Med 65(1):60–63. https://doi.org/10.1055/s-1999-13964.

Zhang L, Ding R, Chai Y, Bonfill M, Moyano E, Oksman-Caldentey KM, Xu T, Pi Y, WangZ ZH, Kai G, Liao Z, Sun X, Tang K (2004) Engineering tropane biosynthetic pathway in *Hyoscyamus niger* hairy root cultures. Proc Natl Acad Sci U S A 101(17):6786–6791. https://doi.org/10.1073/pnas.0401391101

Zhang L, Yang B, Lu B, Kai G, Wang Z, Xia Y, Ding R, Zhang H, Sun X, Chen W, Tang K (2007) Tropane alkaloids production in transgenic *Hyoscyamus niger* hairy root cultures over-expressing putrescine N-methyltransferase is methyl jasmonate-dependent. Planta 225 (4):887–896. https://doi.org/10.1007/s00425-006-0402-1

Zhao J, Davis LC, Verpoorte R (2005) Elicitor signal transduction leading to production of plant secondary metabolites. Biotechnol Adv 23(4):283 333. https://doi.org/10.1016/j.biotechadv.2005.01.003

Zhao K, Zeng J, Zhao T, Zhang H, Qiu F, YangC ZL, LiuX CM, LanX LZ (2017) Enhancing tropane alkaloid production based on the functional identification of Tropine-forming reductase in *Scopolialurida*, a Tibetan medicinal plant. Front Plant Sci 8:1745. https://doi.org/10.3389/fpls.2017.01745

Production of Tropane Alkaloids (TAs) in Plants and In Vitro Cultures of Different Ploidy Levels

6

Ishmael Dehghan and Elnaz Ghotbi

Abstract

Owning multiple pharmaceutical functions, plant-derived tropane alkaloids, especially scopolamine, have been widely used in medicine. Significant progress has been made in the production of tropane alkaloids during the past decades via breeding and cultivation of *Duboisia* cultivates, which are rich in scopolamine. However, unpredicted effects of environmental conditions and the progressive impact of global climate change are two major challenges for sustainable production of these compounds in the near future. In vitro cultures offer a powerful alternative source of TA production which address the uncontrolled effects of environmental variables under field cultivation and most importantly ease manipulation of the culture systems to maximize productivity. However, low quality or quantity of certain TA is still one of the biggest challenges that limit applicability of in vitro cultures. As an evolutionary phenomenon, polyploidy has been known to impact almost all characteristics of a plant species including pharmaceutically important traits such as biomass production and secondary metabolism. However, application of polyploidy has been underestimated and unexplored in production of tropane alkaloids. Here, we review recent progresses in the application of polyploidy in improving the quantity and quality of TA production in plants and in vitro cultures of TA-producing plant species. We also discuss the future directions of the polyploidy research in this field.

I. Dehghan (✉)
Department of Molecular Biology, University of Texas Southwestern Medical Center, Dallas, TX, USA
e-mail: ishmael.dehghan@utsouthwestern.edu

E. Ghotbi
Department of Dermatology, University of Texas Southwestern Medical Center, Dallas, TX, USA

V. Srivastava et al. (eds.), *Tropane Alkaloids*,
https://doi.org/10.1007/978-981-33-4535-5_6

Keywords

Tropane alkaloids · Polyploidy · Tetraploidy · Hairy root cultures · Scopolamine

6.1 Introduction

For thousands of years, human beings used and interacted with plants as the major source of food and medicine. Possessing the biggest genomes in living organisms, and being exposed to various biotic and abiotic stressors, plants evolved many protective mechanisms to survive harsh conditions (Pellicer et al. 2010; Butler 2004; Newman and Cragg 2012). Expanding various metabolic by-products, well-known as secondary metabolites, is one of these defense strategies. Interestingly, many of these by-products are biologically active and important in the interaction of plants with other organisms and have been wisely recognized by human being as useful source of pharmaceuticals (Butler 2004; Newman and Cragg 2012; Wink 2015). As a subcategory of plant alkaloids, TAs, such as atropine, hyoscyamine, and scopolamine (also known as hyoscine), have been widely used as anticholinergic drugs (Kohnen-Johannsen and Kayser 2019). Commercial production of these drugs is limited to the extraction and purification from various solanaceous plants belonging to the genera *Atropa*, *Duboisia*, *Datura*, and *Hyoscyamus* (Mateus et al. 2000; Oksman-Caldentey and Arroo 2000). Although all TAs have the same basic structure, they offer different pharmacological properties. Hyoscyamine is the main compound in many of these plants such as *Hyoscyamus muticus* and *Atropa belladonna*, while scopolamine is only produced in small quantities (Mateus et al. 2000; Oksman-Caldentey and Arroo 2000). Scopolamine has the largest legitimate market among TAs due to its wide application in treatment of nausea, vomiting, motion sickness, as well as smooth muscle spasms (Kohnen-Johannsen and Kayser 2019). Therefore, improving quantity and quality of the TAs in plants and in vitro systems along with other important agricultural traits is critical for sustainable production of scopolamine and other TAs in the future. Manipulation of ploidy level has been suggested as an interesting approach for increasing both the plant biomass accumulation and the production of high value plant secondary metabolites (Lavania 2005). Recent studies have shown some promising benefits of polyploidy in the production of tropane alkaloids. Here, we discuss applicability of chromosome manipulation in TA research area and discuss future direction.

6.2 Polyploidy and Cell Metabolism: Is More Always Better?

Beyond well-established roles in increasing cell size and metabolic output, polyploidy can also induce nonuniform genome and epigenome which leads to a distinct metabolic gene expression and causes metabolome alterations. One simple model of how increased genome can affect transcription suggests a global increase in gene transcription in a simple linear relationship. However, majority of the experimental models disagree a simple linear relationship between chromosome ploidy and transcript levels (Zhang et al. 2010; Coate and Doyle 2010). One major reason can be the extensive epigenetic regulation of gene expression across the genome after changes in DNA dosage. Therefore, the transcription level of the genes related to the pathways of secondary metabolites does not evenly increase across the pathway and can even decrease in some cases.

It is known that increasing volume due to a bigger genome in polyploids decreases surface area/volume ratio, which can consequently change numerous processes such as cell signaling and protein and metabolite trafficking across membrane. It is speculated that these changes in cell volume modulate expression/metabolome changes in polyploid cells (Schoenfelder and Fox 2015).

Duplication of the genome is known to rewire cellular primary and secondary metabolism which can lead to the alteration of the regulatory mechanisms of individual metabolites (Lavania 2005; Dehghan et al. 2012; Ghotbi Ravandi et al. 2013). Redirection of the cell carbon and nitrogen sources from primary to secondary metabolism limits availability of vital primary and constructive elements of the cell growth and maintenance which can potentially slow down plant growth rate (Caretto et al. 2015). Therefore, in some cases, especially in tetraploid clones with very high alkaloid content, a modest decrease in growth rate and overall biomass production is not unusual. In our experience, this is more common in in vitro cultures rather than the plants (Dehghan et al. 2017). The increased chloroplast number and photosynthesis rate can increase the overall biomass and alkaloid production of the induced tetraploid plants. In overall, polyploidy induction can result in the development of new plants with various exciting phenotypes and higher productivity. However, it is likely that in some rare cases, induction of autotetraploid may not be as beneficial as initially expected and even lead to downregulation of a subset of rate-limiting steps in a secondary metabolite pathway or cause restriction of precursors of a biosynthetic pathway which led to a general reduction in total alkaloid yield.

6.3 Polyploidy in TA-Producing Plants

Polyploidy is known as an important driver of plant evolution and specification in natural populations (Adams and Wendel 2005). Likewise, artificial polyploidy has been widely used as an efficient strategy in plant breeding. Many successful cases of polyploid plants with superior features compared to their diploid counterparts have been documented in crops and medicinal plants (Lavania 2005). As one of the most popular form of polyploidy, induction of tetraploidy from diploid plants has been widely used to alter the quality and quantity of various phytopharmaceuticals, including tropane alkaloids (Lavania 2005; Dehghan et al. 2012; Tavan et al. 2015). A common tetraploidy trial consists of the following steps: (1) the induction step in vitro or in vivo mainly by using antimitotic chemicals such as colchicine; (2) regeneration or regrowth phase which can be accelerated by phytohormones; (3) evaluation and confirmation of the tetraploids by various techniques such as karyotyping, flow cytometry, and morphological examinations; (4) establishment of genetically stable tetraploids; and (6) selection for desired phenotypes such as growth rate and biomass production and most importantly quality and quantity of the desired metabolites.

In overall, a general increase in the total content of alkaloids has been observed in various TA-producing species. Pioneer works of Lavania and Srivastava established applicability of induced autopolyploid in improving TA production of various *Hyoscyamus* species (Lavania 1986; Srivastava and Lavania 1990; Lavania and Srivastava 1999). They showed an increase of 16.2%, 22.5%, and 36% in *H. alba*, *niger*, and *muticus*, respectively (Lavania 1986; Srivastava and Lavania 1990; Lavania and Srivastava 1999). Other species such as *Atropa belladonna* and *Datura stramonium* also showed higher yield of TAs (Evans 1989; Dhawan and Lavania 1996). The alkaloid content in tetraploid seeds of *Datura innoxia* and *D. stramonium* was reported to be about twofold higher than that in their diploid seeds (Berkov 2002). In another study, autotetraploid plants of *D. stramonium* were also found to possess higher alkaloid contents than diploid plants (Berkov and Philipov 2002). Autotetraploid hairy root cultures of *D. stramonium* induced by direct transformation of autotetraploid plants produced more alkaloids than the diploid ones (Berkov et al. 2003).

Autotetraploid plants of *H. reticulatus* also showed a sharp increase in scopolamine content, from 0.23 in diploids to 8.66% in the induced tetraploid plants (Madani et al. 2015). Rapid repatterning and genome rearrangements are common features of induced polyploids (Chen and Ni 2006). Therefore, more follow-up experiments are suggested to confirm stability and real alkaloid yield of such newly induced autotetraploids.

Our recent study also demonstrated higher performance of tetraploid *H. muticus* plants in both biomass and scopolamine production (Dehghan et al. 2012). We grew tetraploid plants of *H. muticus* up to seven generations and observed genetic stability and higher performance of these plants over generations (Dehghan et al. 2017). Autotetraploid plants produced a distinct profile of alkaloids, higher amount of scopolamine, and a modest decrease in hyoscyamine production which resulted in a higher scopolamine to hyoscyamine ratio, making these plants pharmaceutically more beneficial. Although the overall scopolamine content of these tetraploid plants is still less than commercially available high scopolamine yield species of *Duboisia*, these tetraploid plants provided a promising starting genetic resource for culture optimization under in vitro conditions and pave the way for more comprehensive studies in *Hyoscyamus* species.

6.4 In Vitro Culture of Tetraploid Transformed Root Cultures for Scopolamine Production

Most of the conventional polyploidy efforts have been studied in vivo with the whole plant system, and the effects of tetraploidy on in vitro cultures have been mainly neglected. Hairy root cultures induced by *Agrobacterium* transformation are potential resource of TA production besides conventional in vivo cultures. A handful of studies have been performed on ploidy manipulation of the hairy root systems. We and others have shown beneficial effects and applicability of tetraploidy in hairy root systems. In a leading study, Pavlov et al. (2009) investigated the biosynthesis of hyoscyamine in *Datura stramonium* hairy roots with different ploidy levels. Although they did not observe a dramatic change in the major component of TAs between diploids and tetraploids, a significant change was observed for the minor compounds. The scopolamine/hyoscyamine ratio of the cultured hairy roots was not reported from the latter study, but the maximal yield of hyoscyamine (177 mg/L) was obtained when hairy roots from tetraploid plants were cultivated in Murashige–Skoog medium supplemented with 6% sucrose (Pavlov et al. 2009). In *H. muticus*, we directly induced hairy roots from diploid and tetraploid parental plants (Dehghan et al. 2012). Stable hairy roots with different ploidy level were established, and a significant change in the growth rate and alkaloid profile of tetraploid clones was observed compared to the diploid counterparts. The studied tetraploid clone showed a lower biomass production but produced more scopolamine than the diploid counterpart under similar growth conditions.

In general, tetraploid clones demonstrated a higher rate of hyoscyamine to scopolamine conversion, a desired phenotype in the field of medicinal plants. We further investigated tetraploid clones of *H. muticus* in a metabolic engineering study. Hairy root cultures overexpressing h6h were established from diploid and tetraploid plants (Dehghan et al. 2017). Among all the diploid and tetraploid clones, the highest level of h6h transgene expression and scopolamine accumulation was interestingly observed in the tetraploid clones. Therefore, tetraploidy provides a better context for metabolic engineering of scopolamine pathway in *H. muticus*. While these results look promising, we concluded that manipulation of ploidy level alone may not be sufficient to increase the level of tropane alkaloid to an economical scale and must be used in combination with other strategies to achieve maximum productivity of TAs.

6.5 Future Directions

In our experience, not all the newly induced tetraploids in one species are the same in terms of alkaloid production. Due to genome rearrangement and genomic restructuring, it is likely that autotetraploids show distinct alkaloid production patterns. Therefore, selection of the tetraploids with desired characteristics is highly important. Tetraploidy can be also used in combination with other strategies to achieve maximum productivity of TAs in plants and root cultures.

One drawback in most of the studies on the induced tetraploid plants is the investigation of the effects of polyploidy in the early generations of induced plants. Evaluation of the genetic and metabolic stability of the induced plants is important for establishing new tetraploid plants with desired traits.

TA biosynthesis and regulation in Solanaceae plants are complicated processes. As a common mechanism, TA biosynthesis takes place in the roots, and the alkaloids are then transported and stored in the aerial parts. Cell-specific compartmentalization of TA biosynthetic proteins also plays important role in the regulation of scopolamine production (Hashimoto et al. 1991; Pramod et al. 2010). It will be interesting to study how ploidy manipulation changes regulation of TA pathway at organ, tissue, and cell levels. Ploidy may affect the cell volume and biochemistry of various cell types, such as pericycle cells vs. aerial cells, differently. Investigation of tetraploidy effects of autotetraploid on the activity of each individual gene and corresponding enzyme in multiple organs and cellular and subcellular compartments at multiple levels and searching for novel regulatory mechanisms at posttranslational modification can be interesting, as well.

Development of metabolomics methods for fast and reliable analysis of the individual metabolites and intermediates of the tropane alkaloid biosynthetic pathway is highly important for wise manipulation of the pathway and genetic engineering of key steps in tetraploids and selection of elite clones. Novel techniques such as single cell transcriptomics, proteomics, and metabolomics on the diploid and tetraploid plants can deepen our understanding of the effects of genome duplication in the TA production pathway.

Another area that is worth being explored is how increased genomic content can protect tetraploids against extreme environmental conditions. The altered physicochemical properties of polyploid plants may contribute to enhanced protective resistance against extreme conditions. The extra copy of genome in induced tetraploids can provide more flexibility against deleterious mutations, as well.

References

Adams KL, Wendel JF (2005) Polyploidy and genome evolution in plants. Curr Opin Plant Biol 8 (2):135–141. https://doi.org/10.1016/j.pbi.2005.01.001

Berkov S (2002) Size and alkaloid content of seeds in induced autotetraploids of Datura stramonium Datura innoxia and Hyoscyamusniger. Pharm Biol 39:329–331

Berkov S, Philipov S (2002) Alkaloid production in diploid and autotetraploid plants of Datura stramonium. Pharm Biol 40(8):617–621

Berkov S, Pavlov A, Kovatcheva P, Stanimirova P, Philipov S (2003) Alkaloid spectrum in diploid and tetraploid hairy root cultures of Datura stramonium. Z Naturforsch 58c:42–46

Butler MS (2004) The role of natural product in chemistry in drug discovery. J Nat Prod 67:2141–2153

Caretto S, Linsalata V, Colella G, Mita G, Lattanzio V (2015) Carbon fluxes between primary metabolism and phenolic pathway in plant tissues under stress. Int J Mol Sci 16 (11):26378–26394. https://doi.org/10.3390/ijms161125967

Chen ZJ, Ni Z (2006) Mechanisms of genomic rearrangements and gene expression changes in plant polyploids. Bioessays 28(3):240–252. https://doi.org/10.1002/bies.20374

Coate JE, Doyle JJ (2010) Quantifying whole transcriptome size, a prerequisite for understanding transcriptome evolution across species: an example from a plant allopolyploid. Genome Biol Evol 2:534–546. https://doi.org/10.1093/gbe/evq038

Dehghan E, Häkkinen ST, Oksman-Caldentey K et al (2012) Production of tropane alkaloids in diploid and tetraploid plants and in vitro hairy root cultures of Egyptian henbane (Hyoscyamusmuticus L.). Plant Cell Tissue Organ Cult 110:35–44. https://doi.org/10.1007/s11240-012-0127-8

Dehghan E, Reed DW, Covello PS et al (2017) Genetically engineered hairy root cultures of Hyoscyamussenecionis and H. muticus: ploidy as a promising parameter in the metabolic engineering of tropane alkaloids. Plant Cell Rep 36:1615–1626. https://doi.org/10.1007/s00299-017-2178-0

Dhawan OP, Lavania UC (1996) Enhancing the productivity of secondary metabolites via induced polyploidy: a review. Euphytica 87:81–89. https://doi.org/10.1007/BF00021879

Evans WC (1989) Trease and evans pharmacognosy. English Language Book Society/Bailliere Tindal, London, pp 93–109

Ghotbi Ravandi E, Rezanejad F, Zolala J, Dehghan E (2013) The effects of chromosome-doubling on selected morphological and phytochemical characteristics of Cichoriumintybus L. J Hor Sci Biotech 88:701–709

Hashimoto A, Hayashi Y, Amano J, Kohno H, Iwanari II, Usudall S, Yamada Y (1991) Hyoscyamine 6 beta-hydroxylase, an enzyme involved in tropane alkaloid biosynthesis, is localized at the pericycle of the root. J Biol Chem 266:4648–4653

Kohnen-Johannsen KL, Kayser O (2019) Tropane alkaloids: chemistry, pharmacology, biosynthesis and production. Molecules 24(4):796. https://doi.org/10.3390/molecules24040796

Lavania UC (1986) Genetic improvement of Egyptian henbane, Hyoscyamusmuticus L. through induced tetraploidy. Theoret Appl Genetics 73:292–298. https://doi.org/10.1007/BF00289288

Lavania UC (2005) Genomic and ploidy manipulation for enhanced production of phyto-pharmaceuticals. Plant Genet Resour 3:170–177. https://doi.org/10.1079/PGR200576

Lavania UC, Srivastava S (1999) Enhanced productivity of tropane alkaloids and fertility in artificial autotetraploids of Hyoscyamus niger L. Euphytica 52:73–77. https://doi.org/10.1007/BF00021317

Madani H, Hosseini B, Dehghan E et al (2015) Enhanced production of scopolamine in induced autotetraploid plants of HyoscyamusreticulatusL. Acta Physiol Plant 37:55. https://doi.org/10.1007/s11738-015-1795-x

Mateus L, Chekaoui S, Christen P, Oksman-Caldentey KM (2000) Simultaneous determination of scopolamine, hyoscyamine and littorine in plants and different hairy root clones of Hyoscyamusmuticus by micellar electrokinetic chromatography. Phytochemistry 54:517–523. https://doi.org/10.1016/S0031-9422(00)00134-5

Newman DJ, Cragg GM (2012) Natural products as sources of new drugs over the 30 years from 1981 to 2010. J Nat Prod 75(3):311–335. https://doi.org/10.1021/np200906s

Oksman-Caldentey KM, Arroo R (2000) Regulation of tropane alkaloid metabolism in plants and plant cell cultures. In: Verpoorte R, Alfermann AW (eds) Metabolic engineering of plant secondary metabolism. Kluwer, Dortrecht, pp 253–281

Pavlov A, Berkov S, Weber J, Bley TH (2009) Hyoscyamine biosynthesis in Datura stramonium hairy root in vitro systems with different ploidy levels. Appl BiochemBiotechnol 157(2):210–225. https://doi.org/10.1007/s12010-008-8264-6

Pellicer J, Fay MF, Leitch IJ (2010) The largest eukaryotic genome of them all? Bot J Linn Soc 164:10–15. https://doi.org/10.1111/j.1095-8339.2010.01072.x

Pramod KK, Singh S, Jayabaskaran C (2010) Expression of hyoscyamine 6 beta-hydroxylase in the root pericycle cells and accumulation of its product scopolamine in leaf and stem tissues of Datura metel L. Plant Sci 178:202–206

Schoenfelder KP, Fox DT (2015) The expanding implications of polyploidy. J Cell Biol 209(4):485–491. https://doi.org/10.1083/jcb.201502016

Srivastava S, Lavania UC (1990) Meiotic regularization, restoration of seed fertility and alkaloid content in the induced autotetraploids of Hyoscyamusalbus L. Plant Breed 104:160–166

Tavan M, Mirjalili MH, Karimzadeh G (2015) In vitro polyploidy induction: changes in morphological, anatomical and phytochemical characteristics of Thymus persicus (Lamiaceae). Plant Cell Tissue Organ Cult 122:573–583. https://doi.org/10.1007/s11240-015-0789-0

Wink M (2015) Modes of action of herbal medicines and plant secondary metabolites. Medicines 2(3):251–286. https://doi.org/10.3390/medicines2030251

Zhang Y, Malone JH, Powell SK et al (2010) Expression in aneuploid Drosophila S2 cells. PLoS Biol 8(2):e1000320. https://doi.org/10.1371/journal.pbio.1000320

Current Progress in Tropane Alkaloid Production by Transgenic and Heterologous Gene Expression Approaches

7

Neeraj Kumar Dubey, Prashant Singh, Ankita Singh, and Satyendra Kumar Yadav

Abstract

The secondary metabolites of nightshade and coca plant families contain important compounds known as tropane alkaloids (TAs). TAs possess pharmaceutical properties and are used in the treatment of several neurological disorders such as cardiac arrhythmia, Parkinson's disease, etc. These compounds are also used as bronchodilators and anesthetics, although their prolonged use has many side effects. Comparing the global consumption rate of the TAs to the production by the original plant source, their yield is too low. Thus, it is important to increase their production. Several TAs such as scopolamine, spermidine, and hyoscyamine have been subjected to the de novo manufacturing processes for their large-scale production. This chapter summarizes the scientific efforts used to scale up the production of TAs for medicinal purposes. Several strategies such as tissue culture, root culture, hairy root culture, and genetically modified microbe (bacteria and yeast)-mediated large-scale production have been described. However, the best strategies are expected to come in near future, which could fulfill the gap of requirement and the production rate of TAs along with their preserved bioactivity.

Keywords

Transgenic plants · Hairy root cultures · Microbial culture · Secondary metabolites · Tropane alkaloids · Scopolamine · Hyoscyamine · Cocaine

N. K. Dubey (✉) · P. Singh · A. Singh
Botany Department, Rashtriya Snatkottar Mahavidyalaya Jamuhai, Jaunpur, UP, India

S. K. Yadav
Zoology Department, Rashtriya Snatkottar Mahavidyalaya Jamuhai, Jaunpur, UP, India

V. Srivastava et al. (eds.), *Tropane Alkaloids*,
https://doi.org/10.1007/978-981-33-4535-5_7

Abbreviations

ABA	Abscisic acid
ASA	Acetylsalicylic acid
H6H	Hyoscyamine 6 β-hydroxylase
HRCs	Hairy root cultures
MeJA	Methyl jasmonate
MS	Murashige and Skoog medium
PMT	Putrescine N-methyltransferase
Ri plasmids	Root-inducing plasmids
SA	Salicylic acid
TAs	Tropane alkaloids
TRI	Tropine-forming reductase

7.1 Introduction

Plants produce different types of secondary metabolites including the phenylpropanoids, isoprenoids, and alkaloids (Hain and Grimmig 2000). These metabolites play important role in different types of interactions between biotic and abiotic stresses (Hain and Grimmig 2000). Among these secondary metabolites, the tropane alkaloids (TAs) are very important for human uses (Kai et al. 2018). TAs are involved in plant defense against herbivores and function as plant's natural protector against other biotic stresses as well (Castillo et al. 2019). TAs are mainly formed in Convolvulaceae, Brassicaceae, Proteaceae, Euphorbiaceae, Erythroxylaceae, Solanaceae, and Moraceae, along with few other plant families (Jirschitzka et al. 2012; Bedewitz et al. 2014; Kohnen-Johannsen and Kayser 2019). Their biosynthetic processes are supposed to be evolved in parallel ways. Therefore, among these families, their biosynthetic pathways may differ from species to species (Jirschitzka et al. 2012; Bedewitz et al. 2014; Kohnen-Johannsen and Kayser 2019; Docimo et al. 2015). Their concentrations may vary between different plants and plant tissues (Lee et al. 2005). Normally, TAs are synthesized in underground plant parts and stored in aerial parts for the plant defense; however, they were reported to be found in higher quantities in roots, followed by aerial parts like stem and leaves (Lee et al. 2005). Metabolic pathways of TA production are well studied in the Solanaceae family (Zhang et al. 2005).

In planta, the biosynthesis of TAs is started by common precursor of amino acids like L-ornithine and L-arginine (Kohnen-Johannsen and Kayser 2019). Different enzymes having reductase, transferase, decarboxylase, and hydrolases activities, like hyoscyamine-6β-hydroxylase (H6H), putrescine N-methyl transferase (PMT), short chain dehydrogenase (SDR), tropine-forming reductase (TRI), pseudotropine-forming reductase (TRII), arginine decarboxylase (ArgDC), diamine oxidase (DAO), and ornithine decarboxylase (OrnDC), are involved in *in planta* production of TAs (Zhang et al. 2005; Yang et al. 2011). Enzymes like PMT and H6H are rate limiting and impart primarily in regulating the ultimate TA pathway flux (Yang et al. 2011). The activities of these enzymes are limited to particular tissues, for example, the *PMT* gene is expressed in the pericycle and the *H6H* gene is expressed in the pericycle as well as in pollen grain and tapetum region (Zhang et al. 2005; Suzuki et al. 1999a; Suzuki et al. 1999b). Being expressed in the pericycle region, these enzymes might mediate in the rapid transportation of root-synthesized TAs to the aerial parts of the plant (Zhang et al. 2005; Suzuki et al. 1999a). Silencing of TA pathway enzymes prevents the TA production, for example, aromatic amino acid aminotransferase (ArAT) silencing which stops scopolamine synthesis (Bedewitz et al. 2014).

More than 200 types of TAs have been reported from plants (Kohnen-Johannsen and Kayser 2019; Gadzikowska and Grynkiewicz 2001). These compounds are chiefly used for many pharmaceutical purposes such as in neurological disorders, toothache, motion sickness, cardiac abnormalities, postoperative nausea, etc. (Bedewitz et al. 2014; Müller 1998; Guggisberg and Hesse 1983). TA interference in neurotransmitter behavior because of the similarity exists in the chemical structure of TAs and acetylcholine, a neurotransmitter molecule. The major sources of TAs are the plants like *Duboisia myoporoides*, *Scopolia lurida*, *Atropa belladonna*, *Datura metel*, *Hyoscyamus muticus*, and *H. niger*, although the TA content may differ in these plants (Zeng et al. 2016; Zhao et al. 2017). All the TAs retain high degree of similarity due to the presence of tropane ring, and they are broadly categorized into groups like cocaine, calystegines, hyoscyamine, and scopolamine (Kohnen-Johannsen and Kayser 2019).

The natural production of TAs is very less compared to the global consumption rate. The natural production of plant TAs is subjected to several biotic and abiotic factors coupled with ecological, climatic, and environmental issues (Kohnen-Johannsen and Kayser 2019; Georgiev et al. 2013). Therefore, a huge gap exists between the production and consumption of plant-derived TAs which leads to the adoption of different strategies for its production in larger quantities (Georgiev et al. 2013). The present chapter summarizes the current progress in the different strategies used for the large-scale production of most studied TAs like cocaine, calystegines, hyoscyamine, and scopolamine (Zhao 2014).

7.2 In Vitro Systems for Large-Scale TA Production.

Several in vitro systems such as cell suspension, shoot and root cultures, and hairy root cultures (HRCs) have been explored to produce various secondary metabolites including TAs. As majority of TAs are produced in plant roots, out of all explored strategies, the HRC-based technique is considered as the most successful method reported (Georgiev et al. 2013; David et al. 1984). However, a number of reports exist where callus and suspension cultures of different plant species have been explored for TA production. The upcoming text briefly describes various in vitro systems for this purpose.

7.2.1 Hairy Root Cultures (HRCs)

The Ri plasmid (hairy root-inducing plasmids) of *Agrobacterium rhizogenes* induces the formation of hairy roots which are subsequently used to establish hairy root cultures (HRCs) for the production of secondary metabolites including the TAs (Georgiev et al. 2008; White and Nester 1980). At the time of transformation, the T-DNA present on bacterial Ri plasmids gets transferred to the host through infection resulting in the formation of transient plants expressing rootlike structures. These rootlike structures are called hairy roots that have many important attributes like fast growth, frequent branching, and huge root mass production in a very short time (David et al. 1984). These hairy roots are also known to produce the desired product normally in higher concentration compared to the non-transformed tissues (Georgiev et al. 2013; Sevon and Oksman-Caldentey 2002). Besides many advantages, the HRCs also have their own limitations, like low production and requirement of huge mass production of required tissues (Mehrotra et al. 2016). Nevertheless, these tissues have more TAs compared to suspension cultures, and it is very easy to establish HRCs with very less input (Tabata et al. 1972; Sheludko and Gerasymenko 2013). With reference to TA production under in vitro conditions, generally, the young stem and leaf explants of the solanaceous plants are used to deliberately infect with *Agrobacterium* strains and generate respective hairy roots. Several Solanaceae plant genera with different species are utilized for the in vitro production of TAs by using their HRCs, for example, *Atropa* (Yang et al. 2011), *Anisodus* (Kai et al. 2011a), *Scopolia* (Zhao et al. 2017), *Duboisia hybrid* (Palazon et al. 2003), *Brugmansia* (Cardillo et al. 2010), *Datura* (Gontier et al. 1994), *Hyoscyamus* (Yamada and Hashimoto 1982), and *Calystegia* (Scholl et al. 2001) (Table 7.1). A detailed account on hairy root-based plant TA production and application of various yield enhancement strategies has been given by Kai et al. (2018).

Different strategies have also been utilized for normal or enhanced production of TAs through HRCs. For example, biotransformation of exogenously supplemented precursor compound to HRCs has also been done by different workers (Hashimoto and Yamada 1983; Srivastava et al. 2017). Hyoscyamine-fed HRCs of *H. niger* resulted into the formation of scopolamine, which is one of the best examples of HRC-mediated biotransformation (Hashimoto and Yamada 1983; Srivastava et al. 2017; Hashimoto et al. 1986). Elicitation is also considered as one of the most talked about strategies for TA production through HRCs of TA-producing plants. A range of abiotic (salicylic acid, $AgNo_3$, $CaCl_2$, $CdCl_2$, MJ, Ag+, iron oxide nanoparticles, and sodium nitroprusside) elicitors have been used to explore TA productivity of hairy root cultures of different species of *Brugmansia*, *Hyoscyamus*, and *Anisodus*. In majority of studies, the use of these elicitors enhanced the overall productivity of native TAs. Similarly, the effect of bacterial elicitors has also been studied in species of *Datura* and *Scopolia* (Srivastava et al. 2017) which revealed a significant of biotic elicitors on TA production through HRCs of TA-producing plants. Hairy roots have also been used to express the genes related to the TA production. During the transgenic expression, strategies like overexpression of native gene (Kai et al.

Table 7.1 Relevant examples of in vitro production of tropane alkaloid production

S. No	System	Strategies/gene utilized (**metabolic engineering*)	Plants	Metabolite (TA)	References
1	Shoot culture	–	*Erythroxylum coca*	Cocaine	Lydon et al. (1993)
2	Root culture	*NtPMT**	*Atropa belladonna*	Calystegines	Rothe et al. (2003)
3	Root culture	–	*A. belladonna*	Calystegines	Rothe et al. (2001))
4	Hairy root cultures (HRCs)	*Hyoscyamine 6β-hydroxylase (HnH6H)**	*A. belladonna*	Scopolamine	Hashimoto et al. (1993)
5	HRCs	–	*Duboisia myoporoides*	Scopolamine	Yukimune et al. (1994)
6	HRCs	*HnH6H**	*Hyoscyamus muticus*	Scopolamine	Jouhikainen et al. (1999)
7	HRCs	*Putrescine-N-methyltransferase* (*NtPMT*)*	*Datura metel* and *H. muticus*	Hyoscyamine and scopolamine	Moyano et al. (2003)
8	HRCs	*HnPMT* and *HnH6H**	*H. niger*	Scopolamine	Zhang et al. (2004)
9	HRCs	*Tropinone reductases* (*DsTR*)*	*A. belladonna*	Hyoscyamine, scopolamine, and calystegine	Richter et al. (2005)
10	HRCs	*NtPMT**	*H. niger*	Scopolamine	Zhang et al. (2007)
11	HRCs	*Tetraploid*	*D. stramonium*	Hyoscyamine	Pavlov et al. (2009))
12	HRCs	*HnPMT* and *HnH6H**	*A. belladonna* L.	Hyoscyamine and scopolamine	Yang et al. (2011)
13	HRCs	*AaPMT* and *AaTRI**	*Anisodus acutangulus*	Scopolamine and anisodine	Kai et al. (2011a)
14	HRCs	*Short chain dehydrogenase* (*SDR*)*	*Scopolia lurida*	Hyoscyamine	Zhao et al. (2017)

(continued)

Table 7.1 (continued)

S. No	System	Strategies/gene utilized (**metabolic engineering*)	Plants	Metabolite (TA)	References
15	HRCs	*HnH6H**	*Duboisia hybrid*	Scopolamine	Palazon et al. (2003)
16	HRCs	*NtPMT**	*S. parviflora*	Scopolamine and hyoscyamine	Lee et al. (2005)
17	HRCs	*HnH6H**	,	Scopolamine	Kang et al. (2005)
18	HRCs	*HnH6H**	*A. baetica*	Scopolamine	Zarate et al. (2006)
19	HRCs	*SpPMT* and *SpH6H**	*S. parviflora*	Scopolamine and hyoscyamine	Kang et al. (2011)
20	HRCs	*NtPMT**	*A. belladonna*	Hyoscyamine	Sato et al. (2001)
21	HRCs	Elicitation—*salicylic acid (SA), acetylsalicylic acid (ASA)*, or *methyl jasmonate* (*MeJA*)	*A. baetica*	Scopolamine	El Jaber-Vazdekis et al. (2008)
22	HRCs	*Upscaling in 1.5 L stirred tank*	*Brugmansia candida*	Scopolamine and anisodamine	Cardillo et al. (2010)
23	HRCs	–	*H. muticus*	Hyoscyamine, scopolamine, and calystegines	(Sevón et al. (1997)
24	HRCs	–	*Calystegia sepium*	Calystegines	Scholl et al., (2001)
25	Transgenic	*AbH6H**	*A. belladonna*	Scopolamine	Suzuki et al. (1999b)
26	Transgenic	*HnTRI* and *HnH6H**	*N. tabacum*	Acetylated forms of tropine	Rocha et al. (2002)
27	Suspension culture	–	*D. innoxia*	Scopolamine and hyoscyamine	Gontier et al. (1994)
28	Callus culture	–	*H. niger*	Hyoscyamine	Yamada and Hashimoto (1982)

29	Callus culture	–	*D. metel*	Hyoscyamine	El-Rahman et al. (2008)
30	Callus culture	–	*E. coca*	Cocaine	Asano (1981)
31	*Saccharomyces cerevisiae*	*BcH6H**	–	Hyoscyamine	Cardillo et al. (2008)
32	*Escherichia coli*	*AaH6H**	–	Scopolamine	Kai et al. (2011b)

2011a), expression of gene in heterologous system (Kohnen-Johannsen and Kayser 2019), or suppression of the competitive pathways (Srivastava et al. 2017) have been aimed.

7.2.2 Suspensions and Callus Cultures

Callus and suspension cultures of TA-producing plants have also been established for TA production (Koul et al. 1983). The cell suspension cultures have the potential to produce huge biomass within a short time. For example, more than 100 g dry weight was obtained within 13 days from *H. muticus* suspension cultures (Koul et al. 1983). In a study, the suspension cultures of *D. innoxia* are also made to yield the TAs, and these suspension cells were subjected to calcium alginate immobilization which resulted in enhanced yields (Gontier et al. 1994). Such studies have suggested the suitability of cell and suspension cultures for TA production. However, molecular instability with respect to the changes in the culturing conditions and long-term maintenance makes this technique somewhat less usable (Mehrotra et al. 2016). Further, due to slow-growing and time-taking process of plant-derived cultures, the production of TAs in heterologous systems, viz., bacteria and yeast, has also been reported (Kohnen-Johannsen and Kayser 2019) (Table 7.1). Compared to the plant-based cell cultures, the microbial cultures such as the bacterial (e.g., *Escherichia coli*) and yeast (e.g., *Saccharomyces cerevisiae*) cultures are grown in shorter time and are easy to handle (Kohnen-Johannsen and Kayser 2019).

7.2.3 Bioreactors

Low production of TAs at commercial level is one of the major limitations, and there is urgent requirement to develop an optimized culture system for TA production (Mehrotra et al. 2016). One of the prominent efforts to deal with this limitation of low productivity includes bioreactor upscaling (Mehrotra et al. 2016). Bioreactor strategy includes scale-up of an in vitro culture of lab-scale bioprocess to larger culture vessels of commercial scale to produce larger amounts of product (Srivastava 2015). Generally, three types of bioreactors, namely, liquid, gas, and hybrid type, have been reported (Kim et al. 2002). Expensive secondary metabolites including TAs can become cheaper by interventions of bioreactor strategy (Cardillo et al. 2010; Giddings et al. 2000). Many bioreactors with automated system have been developed to scale up the production of TAs independently to time and seasonal barriers (Gerth et al. 2007). Many plant-derived cultures are grown in different types of bioreactors for the generation of large amount of secondary metabolites including TAs. For example the HRCs of *Salvia sclarea* grown in shake flasks and bioreactors showed diterpenoid accumulation during the methyl jasmonate (MeJA) treatment (Kuźma et al. 2009) (Table 7.2). Similarly, the root cultures (normal plant roots) and HRCs of *H. niger* were grown in different culture systems such as flasks, stirred tank, mist, and unified mist system, and it was shown that TA contents including

Table 7.2 Bioreactor strategies for TA production

S. no.	Type of culture	Plant name	Tropane alkaloid	References
1	Root culture	*Datura stramonium*	Hyoscyamine and scopolamine content increases	Maldonado-Mendoza et al. (1992)
3	Root culture	*Hyoscyamus niger*	Scopolamine/hyoscyamine content increases	Woo et al. (1996)
4	Adventitious root culture	*Scopolia parviflora*	Scopolamine and hyoscyamine rapidly increase	Min et al. (2007)
2	Hairy root cultures (HRCs)	*Datura innoxia*	,	Boitel-Conti et al. (1995)
5	HRCs	Tobacco expressing H6H	Converted exogenously supplied hyoscyamine to scopolamine	Moyano et al. (2007)
6	HRCs	*Salvia sclarea*	Diterpenoid accumulation	Kuźma et al. (2009)
7	HRCs	*Atropa belladonna*	Increasing hyoscyamine/ scopolamine ratio	Chashmi et al. (2010)
8	HRCs	*Przewalskia tangutica*	Scopolamine/hyoscyamine content increases	Lan and Quan (2010)
9	HRCs	*Brugmansia candida*	Scopolamine and anisodamine scale-up	Cardillo et al. (2010)
10	HRCs	*H. niger*	TA content increases compared to control	Zhao (2014)
11	HRCs	*H. niger*	Scopolamine, hyoscyamine, and cuscohygrine content increases	Jaremicz et al. (2014)

scopolamine were increased more than 20 times within 25 days (Woo et al. 1996) (Table 7.2). Oxygen is used as the main component during the cellular respiration, and it is one of the most critical components during the culturing of plant cells in bioreactor (Zhao 2014; Jaremicz et al. 2014). To ensure proper oxygen supply, small-scale bubble column-type bioreactor was developed, and it was shown that the TA production was increased compared to the control conditions (Jaremicz et al. 2014). Further, the HRCs of *H. niger* grown in bioreactor with better oxygen supply through microbubbles and hybrid bubble-column/spray technique gave manyfold TA production than control conditions (Georgiev et al. 2008; Jaremicz et al. 2014) (Table 7.2). The bioreactor techniques have also been applied for culturing of *Datura stramonium* and *D. innoxia* (Maldonado-Mendoza et al. 1992; Boitel-Conti et al. 1995). Root and HRCs of *D. stramonium* and *D. innoxia*, supplemented with indolebutyric acid and Tween20, respectively, not only helped in the large-scale production of the TAs but also increased their permeabilization in bioreactors (Boitel-Conti et al. 1995; Maldonado-Mendoza et al. 1993) (Table 7.2). Adventitious root cultures of *S. parviflora* grown in bioreactor having small-scale bubble column showed large-scale production of scopolamine and hyoscyamine compared to the flask-type cultures. Further elicitation of these cultures with bacterium

Staphylococcus aureus increased the TA content (Min et al. 2007). These bioreactors also work as biotransformation platforms like the HRCs of *H6H*-expressing tobacco grown in bioreactor, having 5 L size with turbine stirred tank reactor, transforming the exogenously supplied hyoscyamine to scopolamine (Moyano et al. 2007) (Table 7.2). The HRCs and in vitro cultured *A. belladonna* in bioreactors of 100 ml flask size without hormone improved the formation of hyoscyamine/scopolamine ratio when supplemented with nitrate concentration (Chashmi et al. 2010) (Table 7.2). The HRCs of *Przewalskia tangutica* and *B. candida* grown in different bioreactor of 1.5 L stirred tank size, or 5-week culture in MS liquid medium, produced scopolamine and hyoscyamine at higher levels compared to the control plants (Cardillo et al. 2010; Lan and Quan 2010) (Table 7.2).

7.3 In Vitro Production of some Important Plant TAs

7.3.1 Cocaine

Cocaine is used as a psychoactive drug compound. The unlawful cultivation of cocaine-producing plants (normally the members of Erythroxylaceae) and smuggling of cocaine are reported worldwide. There are some reports that erythroxylacean plants are being cultivated in the south of Mexico as an alternative to its recreational product cocaine (Restrepo et al. 2019). There are two species, *Erythroxylum coca* and *Erythroxylum novogranatense*, and four varieties which are cultivated across the world (Restrepo et al. 2019). An alternative source of cocaine is *Boliviana negra*. It is also known as supercoca (a form of *E. coca*), which was developed by the selective breeding strategy and is resistant to herbicides. Cocaine are produced and found in maximum amount in the buds, leaves, embryos, and endosperm of the mature fruit of *E. coca* (Johnson and Elsohly 1991; Docimo et al. 2012). The production and formation of cocaine in plant tissues is affected by the day and night duration as well (Johnson 1993). Several strategies were applied to improve in vitro production of cocaine. Among them, the callus culture was prepared by using green stem and embryo for cocaine production (Asano 1981; Lydon et al. 1993) (Table 7.1). However, the cocaine obtained from such cultures was in lesser amount compared to the parent plants of *E. coca* (Asano 1981; Lydon et al. 1993) (Table 7.1). Recently, Docimo et al. (2015) demonstrated the cocaine production by the callus culture of *E. coca* leaves. The phytohormones and different media also affect the cocaine production significantly (Docimo et al. 2015) (Table 7.1). Like the uses of Anderson's Rhododendron media, Gamborg B5, and modified Murashige and Tuckermedia help in the production of cocaine in in vitro conditions (Docimo et al. 2015). Phytohormones like 2,4-D, indole butyric acid and benzylaminopurine affect the biosynthesis of cocaine in callus culture of *E. coca* (Docimo et al. 2015) (Table 7.1).

7.3.2 Calystegines

Calystegines, the nortropane alkaloid derived from the tropane biosynthetic pathway, are also known as glycosidase inhibitors available in a number of plant species (Table 7.1) (Richter et al. 2007). Calystegine are found in various solanaceous foods, in particular in potatoes and eggplants. Calystegines production and accumulation has been observed in transgenic *Solanum tuberosum* (Lydon et al. 1993). The non-transgenic root culture of *A. belladonna* showed the production of calystegine (Rothe et al. 2003). In another study, these roots in response to high sucrose also exhibited high calystegines; however, presence of other elicitors like chitosan, abscisic acid (ABA), and methyl jasmonate (MeJA) did not significantly affect calystegine production. This has suggested the role of sucrose-related metabolism in calystegine biosynthesis (Rothe et al. 2001). Calystegine production is also reported in hairy roots of different plants. HRCs of *H. muticus* showed the production of calystegines along with hyoscyamine and scopolamine (Sevón et al. 1997). Similarly, the HRCs of *A. belladonna* expressing tropinone reductases showed higher production of calystegine along with hyoscyamine and scopolamine compared to the HRCs of wild-type control (Richter et al. 2005). Further, another report showed the production of calystegines of about 1.5 mg/g dry mass of hairy roots of *Calystegia sepium* (Scholl et al. 2001).

7.3.3 Hyoscyamine and Scopolamine

Currently hyoscyamine and scopolamine are produced at large scale by hybrids of *Duboisia* plants (Kohnen et al. 2018). Tissue culture strategies have been widely applied to produce hyoscyamine and scopolamine (El-Rahman et al. 2008). More than 15 mg/g of scopolamine produced by callus cultures were obtained from different parts of *D. metel* (El-Rahman et al. 2008) (Table 7.1). Further production of hyoscyamine and scopolamine was also performed in the callus culture of *D. leichhardtii* and *D. innoxia* which showed manyfold increased production of these metabolites as compared to the non-cultured tissue (Gontier et al. 1994; El-Rahman et al. 2008).

Hyoscyamine and scopolamine have also been produced by the adventitious root cultures of *H. niger* and *A. belladonna* L. (Kawamura et al. 1996). Apart from TA production from callus and adventitious root cultures, several hairy root cultures of TA-producing plants have also been utilized by expressing different genes related to TA biosynthesis for their enhanced production (Hashimoto et al. 1993; Yukimune et al. 1994; Kang et al. 2011). Though the HRCs of different *Atropa* species have been effectively used for TA production, the development of *A. belladonna* transgenic HRCs expressing *PMT*, *TR,* and *H6H* further demonstrated enhanced production of TAs in comparison with the HRCs of wild-type plants (Yang et al. 2011; Richter et al. 2005; Zarate et al. 2006; Sato et al. 2001).

The HRCs of *D. metel* and *H. muticus*, overexpressing *PMT*, showed increment in hyoscyamine and scopolamine production (Moyano et al. 2003). Further, combining the elicitation and gene engineering strategies in hairy roots of TA-producing plants has given interesting results. Application of SA, ASA, or MeJA to HRCs of *A. baetica* expressing H6H showed about ninefold induction in the production of scopolamine (Zarate et al. 2006; El Jaber-Vazdekis et al. 2008). The HRCs of *Duboisia leichhardtii* and different hybrids of *Duboisia* expressing *H6H* of *H. niger* also showed increment in the production of hyoscyamine and scopolamine compared to the non-cultured plant material (Palazon et al. 2003). The HRCs of *Duboisia myoporoides* showed about 3.5% increment in the production of scopolamine compared to the dry weight of native plant (Yukimune et al. 1994).

HRCs of different species of *Scopolia*, viz., *S. lurida* (Zhao et al. 2017) and *S. parviflora* (Lee et al. 2005), are also used as source of hyoscyamine and scopolamine production. The HRCs of *S. lurida* overexpressing short chain dehydrogenase (*SDR*) gene (Zhao et al. 2017) and *S. parviflora* expressing tobacco *PMT* (Lee et al. 2005) and *S. parviflora H6H* and *PMT* (Kang et al. 2011; Kang et al. 2005) showed more production of hyoscyamine and scopolamine compared to the native plant material. The HRCs of *H. muticus* overexpressing *H6H* (Jouhikainen et al. 1999), *PMT* and *H6H* in *H. niger* (Zhang et al. 2004; Zhang et al. 2007), and callus culture of *H. niger* (Yamada & Hashimoto 1982) showed manyfold increment in the production of hyoscyamine and scopolamine than wild type. The HRCs of *A. acutangulus*, overexpressing *PMT* and *TRI*, showed more yield of TAs compared to the wild-type plants (Kai et al. 2011a). Similarly, the HRCs of *B. candida* growing in 1.5 L stirred tank showed scaled-up production of scopolamine and anisodamine (Cardillo et al. 2010). The transgenic lines of *N. tabacum* expressing *TR* and *H6H* of *H. niger* showed production of acetylated forms of tropine (Rocha et al. 2002).

Apart from regulating the expression level of TA biosynthetic genes, the alteration of ploidy levels of in vitro cultures also affected TA production. The HRCs of tetraploid *D. stramonium* showed about 24.7% more production of hyoscyamine compared to the native plants (Pavlov et al. 2009).

Not only plants but several microbial organisms such as bacteria and yeast were also modified to produce the TAs. In this reference, yeast (*S. cerevisiae*) strain was transformed by expressing *H6H* of *B. candida* and 15 additional genes related to the TA metabolism (Cardillo et al. 2008; Srinivasan and Smolke 2019). The transformed yeast cells produced large amount of hyoscyamine compared to the non-transformed cells (Cardillo et al. 2008; Srinivasan and Smolke 2019) (Table 7.1). Similarly, *E. coli* expressing *A. acutangulus H6H* showed scopolamine production (Kai et al. 2011b).

7.4 Factor Affecting the In Vitro Production of TAs

In vitro production of TAs is affected by several biotic and abiotic factors. Basal culture media, micro- and macronutrients, sucrose content, type and size of culture vessel, ploidy level of explants, cultivation periods, etc. affect the in vitro production of TAs. The TA concentration can be increased up to 200 times in culture of tetraploid plant tissue (Dehghan et al. 2012). Further, effect of calcium compound including calcium alginate on TA production has been observed in *D. innoxia* suspension culture (Gontier et al. 1994). 10 mM calcium chloride supplement in cell culture induced about tenfold more production of scopolamine and hyoscyamine (Gontier et al. 1994). MeJA treatment in hairy root cultures of *H. niger* induces expression of *PMT* and increases the production of scopolamine (Zhang et al. 2007). Increment in sucrose concentration of about 5% increases the production of calystegine in transgenic *A. belladonna* root culture expressing the *N. tabacum*, *PMT* gene (Rothe et al. 2001). Some chemicals like salicylic acid (SA) and acetylsalicylic acid (ASA) induce the synthesis of hyoscyamine in hairy root culture of Algerian *Datura* species (Harfi et al. 2018). Shake flask type, inoculum density, compact growth of cell, continuous illumination, etc. also affect the TA production in in vitro cultures (Gontier et al. 1994; Yamada and Hashimoto 1982; Woo et al. 1996). Different phytohormones such as GA, auxin, NAA, and BAP affect the content of TAs in cultured tissues (Georgiev et al. 2013). Culture time period is also an important parameter for TA production in vitro. For example, in case of about 5-year-long culturing of hairy root lines of *D. stramonium*, positive changes in TA production have been observed (Maldonado-Mendoza et al. 1993). In case of bioreactor upscaling, size and design of culture vessels affects TA synthesis and accumulation most. Reactor vessel with minimized shear stress and optimized aeration leads to the desired production of TAs in bioreactor. Further, different bioreactor probes like pH, temperature controlling these physical and chemical factors, concentration gradients, turbidity, flow rates, and oxygen amount also regulate the TA production. Growing tissues of hairy roots and other culture systems in liquid medium also face stress factor called as hydrodynamic stress, which gives extra burden on the cells due to continuous agitation. This represents another limitation for the TAs production in large culture vessels including bioreactor (Mehrotra et al. 2016). Further, the root hairs also cause problem in mass transport. In a study, the use of computational fluid dynamics (CFD) system in the root cultures of *H. muticus* provides better management in mass transport (Carvalho and Curtis 1998). To avoid clump formation and reduce mass transfer limitation, the use of porous polypropylene membrane tubing system provided better aeration system in the *A. belladonna* root cultures (Kanokwaree and Doran 1998).

7.5 Conclusion

Plants TAs are important naturally occurring metabolites of high pharmaceutical importance. Owing to the drawbacks of direct plant destruction and costly chemical synthesis, in vitro strategies have been adopted for the production of plant TAs in order to reduce the gap between demand and supply. Various in vitro strategies such as callus and suspension cultures and adventitious and hairy root cultures have been explored for optimized TA production. Although several of the reports claimed to have best results in terms of high TA yield in different systems, there is a need to optimize a strategy for large-scale TA production with low cost and labor inputs. In this regard, some of the studies combine one or two strategies together to optimize the final metabolic yield. For example, genetically engineered hairy roots of *A. baetica* expressing *H6H* have showed a ninefold increase in scopolamine content after the treatment of SA, acetylsalicylic acid, and/or MeJA. Similarly, bioreactor upscaling of such roots may show some prospective toward better results. In the same way, transformed microbial systems with overexpressing TA-related genes can also be upscaled in large culture vessels to get higher amounts of TAs. However, upscaling in bioreactor or any large culture vessel is a tough task and needs to be optimizing their culture conditions at first. Although the TAs are produced by different bioreactors, either by growing the HRCs or by other culture strategies, they have some limitations such as unstable productivity of suspension culture, problem in mass transfer of the HRCs, and inadequacy of oxygen supply to the growing cells. So, there is an urgent need for the development of such bioreactors which has the potential to overcome these constrains. Further, different chemicals and physical parameters also need to be optimized prior to culture as these parameters influence the large-scale formation of the TAs in vitro. Further, modified microbes like yeast and bacteria and their further uses can reduce both time and cost of TA production compared to plant tissue culture-based TA production.

References

Asano WY (1981) Growth and alkaloid content of *Erythroxylon Coca*. PhD dissertation, University of Hawaii

Bedewitz MA, Góngora-Castillo E, Uebler JB, Gonzales-Vigil E, Wiegert-Rininger KE, Childs KL, Hamilton JP, Vaillancourt B, Yeo YS, Chappell J, Della Penna D (2014) A root-expressed L-phenylalanine: 4-hydroxyphenylpyruvate aminotransferase is required for tropane alkaloid biosynthesis in *Atropa belladonna*. The Plant Cell 26(9):3745–3762. https://doi.org/10.1105/tpc.114.130534

Boitel-Conti M, Gontier E, Laberche JC, Ducrocq C, Sangwan-Norreel BS (1995) Permeabilization of *Daturainnoxia* hairy roots for release of stored tropane alkaloids. Plantamedica 61 (03):287–290. https://doi.org/10.1055/s-2006-958079

Cardillo AB, Talou JR, Giulietti AM (2008) Expression of *Brugmansia candida* Hyoscyamine 6beta-hydroxylase gene in *Saccharomyces cerevisiae* and its potential use as biocatalyst. Microb Cell Factories 7(1):17. https://doi.org/10.1186/1475-2859-7-17

Cardillo AB, Otálvaro AÁ, Busto VD, Talou JR, Velásquez LM, Giulietti AM (2010) Scopolamine, anisodamine and hyoscyamine production by *Brugmansia candida* hairy root cultures in bioreactors. Process Biochem 45(9):1577–1581. https://doi.org/10.1016/j.procbio.2010.06.002

Carvalho EB, Curtis WR (1998) Characterization of fluid flow resistance in root cultures with a convective flow tubular bioreactor. Biotechnol Bioeng 60:375–384. https://doi.org/10.1002/(SICI)1097-0290(19981105)60:3<375::AID-IT15>3.0.CO;2-L

Castillo G, Calahorra-Oliart A, Núñez-Farfán J, Valverde PL, Arroyo J, Cruz LL, Tapia-López R (2019) Selection on tropane alkaloids in native and non-native populations of *Daturastramonium*. Ecol Evol 9(18):10176–10184. https://doi.org/10.1002/ece3.5520

Chashmi NA, Sharifi M, Karimi F, Rahnama H (2010) Differential production of tropane alkaloids in hairy roots and in vitro cultured two accessions of *Atropa belladonna* L. under nitrate treatments. Zeitschriftfür Naturforschung C 65(5–6):373–379. https://doi.org/10.1515/znc-2010-5-609

David C, Chilton MD, Tempé J (1984) Conservation of T-DNA in plants regenerated from hairy root cultures. Bio/Technology 2:73–76. https://doi.org/10.1038/nbt0184-73

Dehghan E, Häkkinen ST, Oksman-Caldentey KM, Ahmadi FS (2012) Production of tropane alkaloids in diploid and tetraploid plants and in vitro hairy root cultures of Egyptian henbane (*Hyoscyamusmuticus* L.). Plant Cell, Tissue and Organ Culture (PCTOC) 110(1):35–44. https://doi.org/10.1007/s11240-012-0127-8

Docimo T, Reichelt M, Schneider B, Kai M, Kunert G, Gershenzon J, D'Auria JC (2012) The first step in the biosynthesis of cocaine in *Erythroxylum coca*: the characterization of arginine and ornithine decarboxylases. Plant Mol Biol 78(6):599–615. https://doi.org/10.1007/s11103-012-9886-1

Docimo T, Davis AJ, Luck K, Fellenberg C, Reichelt M, Phillips M, Gershenzon J, D'Auria JC (2015) Influence of medium and elicitors on the production of cocaine, amino acids and phytohormones by *Erythroxylum coca*calli. Plant Cell, Tissue and Organ Culture (PCTOC) 120(3):1061–1075. https://doi.org/10.1007/s11240-014-0660-8

el Jaber-Vazdekis N, Barres ML, Ravelo AG, Zárate R (2008) Effects of elicitors on tropane alkaloids and gene expression in *Atropabaetica* transgenic hairy roots. J Nat Prod 71 (12):2026–2031. https://doi.org/10.1021/np800573j

El-Rahman RA, El-Din E-WH, El-Said AGA, Khlifa HD (2008) Production of scopolamine and hyoscyamine in calli and regenerate cultures of *Daturametel* (L.). J Appl Sci Res 4 (12):1858–1866

Gadzikowska M, Grynkiewicz G (2001) Tropane alkaloids in pharmaceutical and phytochemical analysis. Acta Pol Pharm 58:481–492

Georgiev M, Georgiev V, Weber J, Bley T, Ilieva M, Pavlov A (2008) *Agrobacterium rhizogenes*-mediated genetic transformations: a powerful tool for the production of metabolites. In: Wolf T, Koch J (eds) Genetically modified plants. Nova, Hauppauge, pp 99–126

Georgiev V, Marchev A, Berkov S, Pavlov A (2013) Plant in vitro systems as sources of Tropane alkaloids. In: Ramawat K, Mérillon JM (eds) Natural products. Springer, Berlin, Heidelberg. https://doi.org/10.1007/978-3-642-22144-6_6

Gerth A, Schmidt D, Wilken D (2007) The production of plant secondary metabolites using bioreactors. Acta Hortic 764:95–104. https://doi.org/10.17660/actahortic.2007.764.11

Giddings G, Allison G, Brooks D, Carter A (2000) Transgenic plants as factories for biopharmaceuticals. Nat Biotechnol 18(11):1151–1155. https://doi.org/10.1038/81132

Gontier E, Sangwan BS, Barbotin JN (1994) Effects of calcium, alginate, and calciumalginate immobilization on growth and tropane alkaloid levels of a stable suspension cell line of *Daturainnoxia* mill. Plant Cell Rep 13(9):533–536. https://doi.org/10.1007/BF00232951

Guggisberg A, Hesse M (1983) Putrescine, spermidine, spermine, andrelated polyamine alkaloids. In: Brossi A (ed) The alkaloids. Academic Press, New York

Hain R, Grimmig B (2000) Modification of plant secondary metabolism by genetic engineering. In: Metabolic engineering of plant secondary metabolism. Springer, Dordrecht, pp 217–231. https://doi.org/10.1007/978-94-015-9423-3_11

Harfi B, Khelifi L, Khelifi-Slaoui M, Assaf-Ducrocq C, Gontier E (2018) Tropane alkaloids GC/MS analysis and low dose elicitors' effects on hyoscyamine biosynthetic pathway in hairy roots of *Algerian Datura* species. Sci Rep 8(1):1–8. https://doi.org/10.1038/s41598-018-36625-4

Hashimoto T, Yamada Y (1983) Scopolamine production in suspension cultures and redifferentiated roots of *Hyoscyamusniger*. Planta Med 47(4):195–199. https://doi.org/10.1055/s-2007-969984

Hashimoto T, Yukinune Y, Yamada Y (1986) Tropane alkaloid production in *Hyoscyamus* root cultures. J Plant Physiol 124:61–75. https://doi.org/10.1016/S0176-1617(86)80178-X

Hashimoto T, Yun DJ, Yamada Y (1993) Production of tropane alkaloids in genetically engineered root cultures. Phytochemistry 32(3):713–718. https://doi.org/10.1016/S0031-9422(00)95159-8

Jaremicz Z, Luczkiewicz M, Kokotkiewicz A, Krolicka A, Sowinski P (2014) Production of tropane alkaloids in *Hyoscyamusniger* (black henbane) hairy roots grown in bubble-column and spray bioreactors. Biotechnol Lett 36(4):843–853. https://doi.org/10.1007/s10529-013-1426-9

Jirschitzka J, Schmidt GW, Reichelt M, Schneider B, Gershenzon J, D'Auria JC (2012) Plant tropane alkaloid biosynthesis evolved independently in the Solanaceae and Erythroxylaceae. Proc Natl Acad Sci 109(26):10304–10309. https://doi.org/10.1073/pnas.1200473109

Johnson EL (1993) Diurnal fluctuations of cocaine and potential precursors in leaves of *Erythroxylumcoca*. J Biosci 48(11–12):863–866. https://doi.org/10.1515/znc-1993-11-1208

Johnson EL, Elsohly MA (1991) Content and de novo synthesis of cocaine in embryos and endosperms from fruit of *Erythroxylum coca* lam. Ann Bot 68(5):451–453. https://doi.org/10.1093/oxfordjournals.aob.a088277

Jouhikainen K, Lindgren L, Jokelainen T, Hiltunen R, Teeri TH, Oksman-Caldentey KM (1999) Enhancement of scopolamine production in *Hyoscyamusmuticus* L. hairy root cultures by genetic engineering. Planta 208:545–551. https://doi.org/10.1007/s004250050592

Kai G, Yang S, Luo X, Zhou W, Fu X, Zhang A, Zhang Y, Xiao J (2011a) Co-expression of AaPMT and AaTRI effectively enhances the yields of tropane alkaloids in *Anisodusacutangulus* hairy roots. BMC Biotechnol 11(1):43. https://doi.org/10.1186/1472-6750-11-43

Kai G, Liu Y, Wang X, Yang S, Fu X, Luo X, Liao P (2011b) Functional identification of hyoscyamine-6-hydroxylase from *Anisodusacutangulus* and overproduction of scopolamine in genetically-engineered Escherichia coli. Biotechnol Lett 33:1361–1365. https://doi.org/10.1007/s10529-011-0575-y

Kai G, Zhao W, Shi M, Wang Y (2018) Biotechnological interventions of hairy roots of Tropane alkaloid-bearing plants. In: Hairy roots. Springer, Singapore, pp 71–93. https://doi.org/10.1007/978-981-13-2562-5_4

Kang YM, Lee OS, Jung HY, Kang SM, Lee BH, Karigar C, Prasad T, Bahk JD, Choi MS (2005) Overexpression of hyoscyamine 6b-hydroxylase(h6h) gene and enhanced production of tropane alkaloids in *Scopoliaparviflora* hairy root lines. J Microbiol Biotechnol 15(1):91–98

Kang YM, Park DJ, Min JY, Song HJ, Jeong MJ, Kim YD, Kang SM, Karigar CS, Choi MS (2011) Enhanced production of tropane alkaloids in transgenic *Scopoliaparviflora* hairy root cultures over-expressing putrescine N-methyl transferase (PMT) and hyoscyamine-6bhydroxylase (H6H). In Vitro Cell Dev Biol Plant 47(4):516–524. https://doi.org/10.1007/s11627-011-9367-2

Kanokwaree K, Doran PM (1998) Application of membrane tubing aeration and perfluorocarbon to improve oxygen delivery to hairy root cultures. Biotechnol Prog 14:479–486. https://doi.org/10.1021/bp9800283

Kawamura M, Shigeoka T, Akita M, Kobayashi Y (1996) Newly developed apparatus for inoculating plant organs into large-scale fermentor. J Ferment Bioeng 82(6):618–619. https://doi.org/10.1016/S0922-338X(97)81266-4

Kim Y, Wyslouzil BE, Weathers PJ (2002) Secondary metabolism of hairy root cultures in bioreactors. In Vitro Cell Dev Biol Plant 38:1–10. https://doi.org/10.1079/IVP2001243

Kohnen KL, Sezgin S, Spiteller M, Hagels H, Kayser O (2018) Localization and Organization of Scopolamine Biosynthesis in *Duboisiamyoporoides* R. Br. Plant Cell Physiol 59(1):107–118. https://doi.org/10.1093/pcp/pcx165

Kohnen-Johannsen KL, Kayser O (2019) Tropane alkaloids: chemistry, pharmacology, biosynthesis and production. Molecules 24(4):796. https://doi.org/10.3390/molecules24040796

Koul S, Ahuja A, Grewal S (1983) Growth and alkaloid production in suspension cultures of *Hyoscyamusmuticus* as influenced by various cultural parameters. Plantamedica 47(01):11–16. https://doi.org/10.1055/s-2007-969939

Kuźma Ł, Bruchajzer E, Wysokińska H (2009) Methyl jasmonate effect on diterpenoid accumulation in Salvia sclarea hairy root culture in shake flasks and sprinkle bioreactor. Enzym Microb Technol 44(6–7):406–410. https://doi.org/10.1016/j.enzmictec.2009.01.005

Lan X, Quan H (2010) Hairy root culture of *Przewalskiatangutica* for enhanced production of pharmaceutical tropane alkaloids. J Med Plants Res 4(14):1477–1481. https://doi.org/10.5897/JMPR09.289

Lee OS, Kang YM, Jung HY, Min JY, Kang SM, Karigar CS, Prasad DT, Bahk JD, Choi MS (2005) Enhanced production of tropane alkaloids in *Scopoliaparviflora* by introducing the PMT (putrescine N-methyltransferase) gene. In Vitro Cell Dev Biol Plant 41(2):167–172. https://doi.org/10.1079/IVP2004621

Lydon J, Zimmerman RH, Fordham IM, Lusby WR (1993) Tissue culture and alkaloid production of *Erthroxylum coca* var. coca. J Herbs Spices Med Plants 2(1):3–14. https://doi.org/10.1300/J044v02n01_02

Maldonado-Mendoza IE, Del Ayora-Talavera TR, Loyola-Vargas VM (1992) Tropane alkaloid production in Daturastramonium root cultures. In Vitro–Plant 28(2):67. https://doi.org/10.1007/BF02823021

Maldonado-Mendoza IE, Ayora-Talavera T, Loyola-Vargas VM (1993) Establishment of hairy root cultures of *Daturastramonium*. Characterization and stability of tropane alkaloid production during long periods of subculturing. Plant Cell Tissue Organ Cult 33(3):321–329. https://doi.org/10.1007/BF02319018

Mehrotra S, Mishra S, Srivastava V (2016) Bioreactor technology for hairy roots cultivation. In: Bioprocessing of plant in vitro systems. Reference Series in Phytochemistry. Springer, Cham

Min JY, Jung HY, Kang SM, Kim YD, Kang YM, Park DJ, Prasad DT, Choi MS (2007) Production of tropane alkaloids by small-scale bubble column bioreactor cultures of *Scopoliaparviflora* adventitious roots. Bioresour Technol 98(9):1748–1753. https://doi.org/10.1016/j.biortech.2006.07.033

Moyano E, Jouhikainen K, Tammela P, Palazón J, Cusidó RM, Piñol MT, Teeri TH, Oksman-Caldentey KM (2003) Effect of pmt gene overexpression on tropane alkaloid production in transformed root cultures of *Daturametel* and *Hyoscyamusmuticus*. J Exp Bot 54:203–211. https://doi.org/10.1093/jxb/erg014

Moyano E, Palazón J, Bonfill M, Osuna L, Cusidó RM, Oksman-Caldentey K-M, Piñol MT (2007) Biotransformation of hyoscyamine into scopolamine in transgenic tobacco cell cultures. J Plant Physiol 164:521–524. https://doi.org/10.1016/j.jplph.2006.06.012

Müller JH u L (1998) Ein Beitragzur Kulturgeschichte der Nachtschattengewächse Hexensalben und Liebestränke. Gesnerus Swiss. J Hist Med Sci 55:205–220

Palazon J, Moyano E, Cusido RM, Bonfill M, Oksman-Caldentey KM, Pinol MT (2003) Alkaloid production in *Duboisia* hybrid hairy roots and plants overexpressing the h6h gene. Plant Sci 165 (6):1289–1295. https://doi.org/10.1016/S0168-9452(03)00340-6

Pavlov A, Berkov S, Weber J, Bley T (2009) Hyoscyamine biosynthesis in *Daturastramonium* hairy root in vitro systems with different ploidy levels. Appl Biochem Biotechnol 157 (2):210–225. https://doi.org/10.1007/s12010-008-8264-6

Restrepo DA, Saenz E, Jara-Muñoz OA, Calixto-Botía IF, Rodríguez-Suárez S, Zuleta P, Chavez BG, Sanchez JA, D'Auria JC (2019) Erythroxylum in focus: an interdisciplinary review of an overlooked genus. Molecules 24(20):3788. https://doi.org/10.3390/molecules24203788

Richter U, Rothe G, Fabian AK, Rahfeld B, Dreager B (2005) Overexpression of tropinonereductases alters alkaloid composition in *Atropa belladonna* root cultures. J Exp Bot 56(412):645–652. https://doi.org/10.1093/jxb/eri067

Richter U, Sonnewald U, Dräger B (2007) Calystegines in potatoes with genetically engineered carbohydrate metabolism. J Exp Bot 58(7):1603–1615. https://doi.org/10.1093/jxb/erl295

Rocha P, Stenzel O, Parr A et al (2002) Functional expression of tropinonereductase I (trI) and hyoscyamine-6β-hydroxylase (h6h) from *Hyoscyamusniger* in *Nicotianatabacum*. Plant Sci 162:905–913. https://doi.org/10.1016/S0168-9452(02)00033-X

Rothe G, Garske U, Dräger B (2001) Calystegines in root cultures of *Atropa belladonna* respond to sucrose, not to elicitation. Plant Sci 160(5):1043–1053. https://doi.org/10.1016/S0168-9452(01)00355-7

Rothe G, Hachiya A, Yamada Y, Hashimoto T, Dräger B (2003) Alkaloids in plants and root cultures of *Atropa belladonna* overexpressing putrescine N-methyltransferase. J Exp Bot 54 (390):2065–2070. https://doi.org/10.1093/jxb/erg227

Sato F, Hashimoto T, Hachiya A et al (2001) Metabolic engineering of plant alkaloid biosynthesis. Proc Natl Acad Sci U S A 98:367–372. https://doi.org/10.1073/pnas.98.1.367

Scholl Y, Höke D, Dräger B (2001) Calystegines in *Calystegiasepium* derive from the tropane alkaloid pathway. Phytochemistry 58(6):883–889. https://doi.org/10.1016/S0031-9422(01)00362-4

Sevon N, Oksman-Caldentey KM (2002) *Agrobacterium rhizogenes*-mediated transformation: root cultures as a source of alkaloids. Planta Med 68(10):859–868. https://doi.org/10.1055/s-2002-34924

Sevón N, Dräger B, Hiltunen R, Oksman-Caldentey KM (1997) Characterization of transgenic plants derived from hairy roots of *Hyoscyamusmuticus*. Plant Cell Rep 16(9):605–611. https://doi.org/10.1007/BF01275500

Sheludko Y, Gerasymenko I (2013) Biosynthetic potential of hairy roots for production of new natural products. In: Chandra S et al (eds) Biotechnology for medicinal plants. Springer, Berlin. https://doi.org/10.1007/978-3-642-29974-2_10

Srinivasan P, Smolke CD (2019) Engineering a microbial biosynthesis platform for de novo production of tropane alkaloids. Nat Commun 10(1):1–5. https://doi.org/10.1038/s41467-019-11588-w

Srivastava V (2015) Studies on hairy roots as bioreactors for normal and elicited biotransformation of exogenous molecules. Ph.D thesis, University of Lucknow, India

Srivastava V, Mehrotra S, Mishra S (2017) Biotransformation through hairy roots: perspectives, outcomes, and major challenges. In: Transgenesis and secondary metabolism. Springer, Cham, pp 1–24

Suzuki KI, Yamada Y, Hashimoto T (1999a) Expression of *Atropa belladonna*putrescine N-methyltransferase gene in root pericycle. Plant Cell Physiol 40(3):289–297. https://doi.org/10.1093/oxfordjournals.pcp.a029540

Suzuki KI, Yun DJ, Chen XY, Yamada Y, Hashimoto T (1999b) An *Atropa belladonna* hyoscyamine 6β-hydroxylase gene is differentially expressed in the root pericycle and anthers. Plant Mol Biol 40(1):141–152. https://doi.org/10.1023/A:1026465518112

Tabata M, Yamanoto H, Hiraoka N, Konoshima M (1972) Organization and alkaloid production in tissue cultures of *Scopoliaparviflora*. Phytochemistry 11:949–955. https://doi.org/10.1016/S0031-9422(00)88438-1

White F, Nester EW (1980) Hairy root: plasmid encodes virulence in *Agrobacterium rhizogenes*. J Bacteriol 141:1134–1141

Woo SH, Park JM, Yang JW (1996) Root culture using a mist culture system and estimation of scale-up feasibility. J Chem Technol Biotechnol 65(4):355–362. https://doi.org/10.1002/(SICI)1097-4660(199604)65:4<355::AID-JCTB464>3.0.CO;2-4

Yamada Y, Hashimoto T (1982) Production of tropane alkaloids in cultured cells of *Hyoscyamusniger*. Plant Cell Rep 1(3):101–103. https://doi.org/10.1007/BF00272363

Yang C, Chen M, Zeng L, Zhang L, Liu X, Lan X, Tang K, Liao Z (2011) Improvement of Tropane alkaloids production in hairy root cultures of'*Atropabelladonna* 'by overexpressing pmt and h6h genes. Plant Omics 4(1):29

Yukimune Y, Hara Y, Yamada Y (1994) Tropane alkaloid production in root cultures of *Duboisiamyoporoides* obtained by repeated selection. Biosci Biotechnol Biochem 58(8):1443–1446. https://doi.org/10.1271/bbb.58.1443

Zarate R, El Jaber-Vazdekis N, Medina B, Ravelo AG (2006) Tailoring tropane alkaloid accumulation in transgenic hairy roots of *Atropabaetica* by over-expressing the gene encoding hyoscyamine 6b-hydroxylase. Biotechnol Lett 28(16):1271–1277. https://doi.org/10.1007/s10529-006-9085-8

Zeng J, Zhao K, Lan X, Liao Z (2016) Analyzing the contents of tropane alkaloids in *Scopolialurida*, a resource plant species of Tibetan medicines. Sci Technol Tibet 279:60–62. https://doi.org/10.3732/ajb.1000010

Zhang L, Ding R, Chai Y, Bonfill M, Moyano E, Oksman-Caldentey KM, Xu T, Pi Y, Wang Z, Zhang H, Kai G (2004) Engineering tropane biosynthetic pathway in *Hyoscyamusniger* hairy root cultures. Proc Natl Acad Sci 101(17):6786–6791. https://doi.org/10.1073/pnas.0401391101

Zhang L, Kai GY, Lu BB, Zhang HM, Tang KX, Jiang JH, Chen WS (2005) Metabolic engineering of tropane alkaloid biosynthesis in plants. J Integr Plant Biol 47:136–143. https://doi.org/10.1111/j.1744-7909.2005.00024.x

Zhang L, Yang B, Lu B, Kai G, Wang Z, Xia Y, Ding R, Zhang H, Sun X, Chen W, Tang K (2007) Tropane alkaloids production in transgenic *Hyoscyamusniger* hairy root cultures over-expressing putrescine N-methyltransferase is methyl jasmonate-dependent. Planta 225 (4):887–896. https://doi.org/10.1007/s00425-006-0402-1

Zhao B (2014) Alkaloid production by hairy root cultures. All Graduate Theses and Dissertations 3884. https://digitalcommons.usu.edu/etd/3884

Zhao K, Zeng J, Zhao T, Zhang H, Qiu F, Yang C, Zeng L, Liu X, Chen M, Lan X, Liao Z (2017) Enhancing tropane alkaloid production based on the functional identification of tropine-forming reductase in *Scopolialurida*, a tibetan medicinal plant. Front Plant Sci 8:1745. https://doi.org/10.3389/fpls.2017.01745

Tropane Alkaloid Biosynthesis in Plants: Insights from Transcriptome Analysis

8

Asosii Paul, Bendangchuchang Longchar, and Jeremy Dkhar

Abstract

Tropane alkaloids (TAs) are a special class of alkaloids found naturally in a diverse group of flowering plant families. To date, about 200 TAs are known, the most prominent being hyoscyamine, scopolamine, calystegine, and cocaine. These compounds possess pharmacological properties and are used in medicine as anticholinergic agents and stimulants. Because of their medicinal value, tropane alkaloids have been the subject of study for several years now. Over the years, research has been directed at elucidating the biosynthetic pathways leading to the production of pharmacologically active TAs. The present chapter discusses recent developments in the understanding of TA biosynthesis with emphasis on the genes involved in the TA biosynthetic pathways and the role transcriptome profiling played in their identification. In recent years, mining of the transcriptome data of TA-producing plants, such as *Atropa belladonna*, has led to a near-complete elucidation of the biosynthesis of hyoscyamine and scopolamine. Advances in gene elucidation made through such studies can be potentially used for metabolic engineering in transgenic plant systems or microbial platforms to sustainably meet the global demand of pharmaceutically important TAs.

A. Paul (✉)
Department of Botany, Nagaland University, Lumami, Nagaland, India
e-mail: apaul@nagalanduniversity.ac.in

B. Longchar
Department of Life Science, Pachhunga University College, Mizoram University, Aizawl, Mizoram, India
e-mail: blongchar@pucollege.edu.in

J. Dkhar
AMACIP Division, CSIR-Institute of Himalayan Bioresource Technology, Palampur, Himachal Pradesh, India
e-mail: jeremydkhar@ihbt.res.in

V. Srivastava et al. (eds.), *Tropane Alkaloids*,
https://doi.org/10.1007/978-981-33-4535-5_8

Keywords

Gene expression · Promoter · Secondary metabolites · TA biosynthetic pathways · Transcriptome · Tropane alkaloid

Abbreviations

ADC	Arginine decarboxylase
AIH	Agmatine deiminase
AT4	Aromatic amino acid aminotransferase 4
cDNA	Complementary DNA
CPA	N-carbamoyl putrescine amidase
CS	Cocaine synthase
H6H	Hyoscyamine 6β-hydroxylase
LS	Littorine synthase
MecgoR	Methylecgonone reductase
MeJA	Methyl jasmonate
MPO	N-methylputrescine oxidase
ODC	Ornithine decarboxylase
PMT	Putrescine N-methyltransferase
PPAR	Phenylpyruvic acid reductase
PYKS	Type III polyketide synthase
TA	Tropane alkaloids
TR-I	Tropinone reductase I
TR-II	Tropinone reductase II
UGT1	Phenyllactate UDP-glycosyltransferase
VIGS	Virus-induced gene silencing

8.1 Introduction

Tropane alkaloids (TAs) belong to a special class of alkaloids—a diverse group of nitrogen—containing compounds found mostly in plants but also occur in certain endophytic fungi (Naik et al. 2017) that is made up of a seven-carbon ring attached to a nitrogen bridge (Bedewitz et al. 2014; Kohnen-Johannsen and Kayser 2019). They are more prominently produced in several flowering plant families and are recognized for their role in plant defense (Shonle and Bergelson 2000; Castillo et al. 2014). Tropane alkaloid-containing plants are among the oldest traditional medicines known to man. Since chemical synthesis is difficult and expensive, plants of the genus *Duboisia* are still cultivated for the extraction of pharmaceutically important TAs (Ullrich et al. 2017). Over 200 TAs are known so far: atropine, scopolamine, cocaine, and calystegine represent the most prominent TAs. Atropine and scopolamine serve as anticholinergic agents while cocaine acts as a stimulant (Lounasmaa and Tamminem 1993).

Atropine is the first TA identified in plants. It was isolated from the roots of the solanaceous perennial herb *Atropa belladonna* in 1832 by the German pharmacist HFG Mein (Eich 2008). A year later, PL Geiger and O Hesse isolated atropine from the leaves of *A. belladonna*. In 1833, Geiger isolated hyoscamine from the seeds of another solanaceous annual herb *Hyoscyamus niger*. Geiger also reported the isolation of atropine, which he called "daturine," from another solanaceous annual herb *Datura stramonium*. Several years later, atropine became recognized as the corresponding racemate of hyoscamine, i.e., the 50:50 mixture of *S*-(−)-hyoscyamine and its *R*-(+)-isomer (Eich 2008). Pure hyoscamine (*S*-(−)-isomer) can be extracted from younger organs, while matured organs contained a considerable amount of both forms, i.e., the *S*-(−)- and the *R*-(+)-forms (Eich 2008). In 1855, Friedrich Gaedecke was credited with the isolation of the tropane alkaloid compound cocaine—which he called erythroxyline—from the leaves of *Erythroxylum coca* (Kohnen-Johannsen and Kayser 2019). In 1862, Albert Niemann extracted and purified the compound and named it "cocaine." In 1881, Ladenburg reported the isolation of a novel tropane alkaloid from *H. niger*, which he named as "hyoscine." Schmidt and Henschke isolated an identical compound from the roots of the solanaceous perennial herb *Scopolia japonica* and named it "hyoscine-Schmidt" (Eich 2008). Schmidt later used the name "scopolamine" to refer to the same compound that was isolated from a different herb *S. carniolica*. Scopolamine was later described as the levorotatory isomer of hyoscine (Pearn and Thearle 1982). In addition to the three major TAs, there exist a fourth kind of polyhydroxy alkaloids called calystegines. All calystegines contain a nortropane skeleton with 3–5 hydroxyl groups at various positions (Dräger et al. 1994) and are more widespread as compared to hyoscyamine and scopolamine (Docimo et al. 2012). These compounds were first isolated from the roots of the perennial herb *Calystegia sepium* (Goldmann et al. 1990). Jocković et al. (2013) demonstrated that calystegines can inhibit glucosidase activity by binding to the active sites of two α-glucosidases maltase and sucrase, which in turn can be used for the treatment of diabetes.

The biosynthesis of the pharmacologically active TAs has drawn the attention of many scientists worldwide. The tremendous work put in by early investigators has resulted in the *in planta* dissection of the chemical route of TA biosynthesis via the in vivo feeding of radiolabeled precursors to TA-producing plants such as *D. stramonium* and *A. belladonna* (Docimo et al. 2012; Zhao et al. 2020). Recent developments in TA biosynthesis have led to an almost complete understanding of the biosynthetic pathway, especially the biosynthesis of hyoscyamine and scopolamine in the Solanaceae family. Part of this progress is attributed to recent advances in high-throughput sequencing technologies (RNA-seq). For example, the Michigan State University has generated and released the transcriptome data of *A. belladonna* (http://medicinalplantgenomics.msu.edu/), whereby several novel genes have been identified such as *type III polyketide synthase* (*AbPYKS*) (Bedewitz et al. 2018; Huang et al. 2019), *AbCYP82M3* (Bedewitz et al. 2018), *aromatic amino acid aminotransferase* (*AbAT4*) (Bedewitz et al. 2014), *phenylpyruvic acid reductase*

(*AbPPAR*) (Qiu et al. 2018), *phenyllactate UDP-glycosyltransferase* (*AbUGT1*) (Qiu et al. 2020), and *littorine synthase* (*AbLS*) (Qiu et al. 2020). Similarly, transcriptome analyses have helped in the identification of key metabolic pathway genes in other TA-producing plants including *E. coca* and *H. niger*, such as *methylecgonone reductase* (*EcMecgoR*) (Jirschitzka et al. 2012), *cocaine synthase* (*EcCS*) (Schmidt et al. 2015), *HnCYP80F1* (Li et al. 2006), *putrescine N-methyltransferase* (*HnPMT*; Geng et al. 2018), and *ornithine decarboxylase* (*HnODC*; Zhao et al. 2019). In contrast to hyoscyamine and scopolamine biosynthesis, detailed information on the calystegine and cocaine biosynthesis pathway is lacking, warranting more research.

Many secondary metabolites, including TAs, are synthesized in specific cells/ tissues of the plant. In TA-producing Solanaceae plants, root tissues are the principal site of TA biogenesis (Hashimoto and Yamada 1994; Fodor and Dharanipragada 1990; Facchini 2001). This invariably resulted in the generation and analyses of root-specific transcriptomes (Li et al. 2006, Bedewitz et al. 2014; Huang et al. 2019). Several novel genes of the hyoscyamine and scopolamine biosynthetic pathway were identified from *A. belladonna* and *H. niger* root transcriptomes (Li et al. 2006; Bedewitz et al. 2018; Huang et al. 2019; Qiu et al. 2020). Promoter analysis is an important functional genomics approach to understand pathway gene functions and their regulatory network. Histochemical analyses of promoter activities of TA biosynthesis genes, such as *AbH6H* and *HnPMT*, consolidated the elucidation of the genes as well as the site of TA biosynthesis in *A. belladonna* and *H. niger*, respectively (Suzuki et al. 1999b; Geng et al. 2018).

In the present chapter, a list of TA-producing plant species belonging to a diverse group of flowering plant families is presented and discussed. We also describe in brief the biosynthetic pathways leading to the production of hyoscyamine, scopolamine, calystegines, and cocaine. Here, efforts are made to describe in detail the various TA biosynthetic genes, including those that were identified and elucidated by the use of transcriptome and promoter analysis based on currently available data. The use of transcriptome data generated from TA-producing plants has contributed significantly in the elucidation of the TA biosynthetic pathway, particularly of hyoscyamine and scopolamine biosynthesis, which is now almost completely understood.

8.2 Tropane Alkaloids Occur Naturally in Diverse Flowering Plant Families

Tropane alkaloids occur naturally in 45 plant genera of ten flowering plant families (Bick et al. 1979; Al-Said et al. 1989; Arbain et al. 1991; Lounasmaa and Tamminem 1993; Asano et al. 1994; Griffin and Lin 2000; Dräger 2004; Hanuš et al. 2005; Chang et al. 2006; Backlund 2010; Jirschitzka et al. 2013, Table 8.1). These secondary metabolites have been detected predominantly in the Solanaceae family. Most members of the family contain at least one of the major TAs as the principal constituent. For example, hyoscyamine is the principal TA found in *A. belladonna*.

Table 8.1 Tropane alkaloid-containing plants and the principal TAs present

Sl. no.	Family	Genus	Species	Principal TA	References
1	Proteaceae	*Agastachys*	*A. odorata*	6β-Acetoxy-3α-tigloyloxytropane; 3α-(4-hydroxybenzoyloxy)trop-6-ene	Bick et al. (1979)
		Bellendena	*B. montana*	Bellendine	Griffin and Lin (2000)
		Darlingia	*Darlingia* spp.	Darlingine	
		Knightia	*K. strobilina*	Strobiline	
2	Moraceae	*Morus*	*M. alba*	Calystegines	Asano et al. (1994)
3	Fabaceae	*Colutea*	*C. orientalis*	3α-(4′-Methoxyphenylacetoxy)-7β-hyroxytropane	Lounasmaa and Tamminen (1993)
4	Erythroxylaceae	*Erythroxylum*	*Erythroxylum* spp.	Tropine	Griffin and Lin (2000)
			E. cataractarum	Cuscohygrine	
			E. coca	Cocaine	
			E. novogranatense	Cocaine	
			E. Zambesiacum	3α-(3′,4′,5′-Trimethoxybenzoyl)oxytropane	
			E. hypericifolium (root)	3α-Phenylacetoxynortropan-6β-ol	Al-Said et al. (1989)
			E. hypericifolium (stem)	Hygrine	
5	Phyllanthaceae	*Phyllanthus*	*P. discoides*	Phyllalbine	Jirschitzka et al. (2013)
6	Rhizophoraceae	*Bruguiera*	*Bruguiera* spp.	Brugine	Griffin and Lin (2000)
		Crossostylis	*Crossostylis* spp.	Brugine	
		Pellacalyx	*P. axillaris*	Nortropinyl cinnamate	Arbain et al. (1991)
7	Brassicaceae	*Cochlearia*	*Cochlearia* spp.	Cochlearine	Brock et al. (2006)
8	Olacaceae	*Heisteria*	*H. olivae*	Scopolamine	Backlund (2010)
9	Convolvulaceae	*Astripomoea*	*A. malvacea*	Astrimalvine	Ott et al. (2007)
		Calystegia	*C. sepium*	Calystegines	Griffin and Lin (2000)

(continued)

Table 8.1 (continued)

Sl. no.	Family	Genus	Species	Principal TA	References
		Convolvulus	*Convolvulus* spp.	Convolvine	Griffin and Lin (2000)
		Ericybe	*E. obtusifolia*	Baotongteng	Griffin and Lin (2000)
			E. Elliptilimba	Erycibelline	
			E. hainanensis	Erycibelline	
		Evolvulus	*E. sericeus*	Convolvine	Griffin and Lin (2000)
		Falkia	*F. repens*	Calystegines	Schimming et al. (2005)
		Ipomoea	*I. Calobra*	Calystegines	Griffin and Lin (2000)
			I. velutina	Ipvelutine	Ott et al. (2013)
		Merremia	*Merremia* spp.	Merresectines	Jenett-Siems et al. (2005)
10	Solanaceae	*Anisodus*	*A. tanguticus*	Anisodamine	Chang et al. (2006)
		Anthocercis	*Anthocercis* spp.	Hyoscyamine or scopolamine	Griffin and Lin (2000)
			A. littorea	Littorine and meteloidine	
			A. albicans	Tropine	
		Anthotroche	*Anthotroche* spp.	Hyoscyamine	Griffin and Lin (2000)
		Atropa	*A. belladonna*	Hyoscyamine	Griffin and Lin (2000)
		Brugmansia	*Brugmansia* spp.	Scopolamine and/or hyoscyamine	Griffin and Lin (2000)
		Crenidium	*Crenidium* spp.	Hyoscyamine	Griffin and Lin (2000)
		Cyphanthera	*C. tasmanica*	Scopolamine	El Imam and Evans (1984)
		Datura	*D. Ceratocaula*	Hyoscyamine	Griffin and Lin (2000)
			D. stramonium	Hyoscyamine	
			D. metel	Scopolamine	
			D. wrightii	Scopolamine	
			D. ferox	Scopolamine	
		Duboisia	*Duboisia* spp.	Scopolamine	Griffin and Lin (2000)
			D. myoporoides N	Scopolamine	

			D. myoporoides S	Hyoscyamine	
			D. myoporoides Q	Low amount of TA	
		Grammosolen	*G. dixonii*	Scopolamine	Griffin and Lin (2000)
		Hyoscyamus	*Hyoscyamus* spp.	Hyoscyamine	Griffin and Lin (2000)
		Latua	*Latua* spp.	Hyoscyamine and tigloidine	Griffin and Lin (2000)
		Mandragora	*Mandragora* spp.	Hyoscyamine	Hanuš et al. (2005)
		Nicandra	*Nicandra* spp.	Tropine	Griffin and Lin (2000)
		Physalis	*P. peruviana*	Tigloidine	Griffin and Lin (2000)
		Physochlaina	*Physochlaina* spp.	Hyoscyamine	Griffin and Lin (2000)
		Przewalskia	*Przewalskia* spp.	Hyoscyamine	Griffin and Lin (2000)
		Salpichroa	*S. origanifolia*	Cuscohygrine and pseudotropine	Griffin and Lin (2000)
		Schizanthus	*S. pinnatus*	Schizanthine	Griffin and Lin (2000)
			S. hookeri	Tropine and schizanthine	
			S. grahamii	Schizanthine and grahamine	
			S. tricolor	Schizanthine	Humam et al. (2011)
		Scopolia	*Scopolia* spp.	Hyoscyamine	Griffin and Lin (2000); Zhao et al. (2017)
		Solandra	*Solandra* spp.	Atropine and/or hyoscyamine	Griffin and Lin (2000)
		Solanum	*S. tuberosum*	Calystegines	Dräger (2004)
			S. Lycopersicon	Calystegines	
		Symonanthus	*S. aromaticus*	Scopolamine	Griffin and Lin (2000)
		Withania	*W. somnifera*	Tropine and pseudotropine	Griffin and Lin (2000)

Species of the genus *Datura* contain either hyoscyamine or scopolamine as the principal TA, while most members of the genus *Duboisia* contain scopolamine as the dominant TA. For *Duboisia myoporoides*, however, different varieties show different TA contents. The variety at north of Gosford, New South Wales, Australia, contains scopolamine, while the one at the south of Gosford possesses hyoscyamine as the major alkaloid. A third variety exists in which little or no TA is reported (Griffin and Lin 2000). Besides the major TAs, members of the Solanaceae family contain novel TAs such as anisodamine and littorine, which are key intermediates in hyoscyamine and scopolamine biosynthesis. Anisodamine, which is related to hyoscyamine and scopolamine, was isolated from the Solanaceous herb *Scopolia tangutica* (*Anisodus tanguticus*) in 1975 by Chinese scientists (Chang et al. 2006; Poupko et al. 2007). Most species of *Anthocercis* contain either hyoscyamine or scopolamine, but in *A. littorea* the major TA detected was littorine (Cannon et al. 1969). Cocaine occurs mainly in the cultivated species of the genus *Erythroxylum*, i.e., *E. coca* and *E. novogranatense* of the family Erythroxylaceae. Most species of the genus contain tropine as the major TA. In *E. hypericifolium*, however, different tissue parts show different TA profiles: 3α-phenylacetoxynortropan-6β-ol represent the major TA in the root, while hygrine was detected as the dominant TA in the stem (Al-said et al. 1989). Cochlearine is a 3-hydroxybenzoate ester of tropine, first isolated from *Cochlearia officinalis* of the family Brassicaceae (Liebisch et al. 1973). Brock et al. (2006) conducted further investigation on the presence of cochlearine in other *Cochlearia* species, including different members of the Brassicaceae family. Other than *Cochlearia*, no member of the Brassicaceae family contains cochlearine (Brock et al. 2006). However, a majority of the Brassicaceae members, including *Cochlearia*, contain calystegines. The convolvulaceous genus *Merremia* contains merresectines as the major TA. But not all species of *Merremia* contain tropane alkaloids (Jenett-Siems et al. 2005).

8.3 The Biosynthesis of Tropane Alkaloids Is Almost Completely Understood

The decarboxylation of ornithine to putrescine by the enzyme ornithine decarboxylase (ODC) marks the first step in TA biosynthesis (Michael et al. 1996) (Fig. 8.1). Putrescine can also be produced through a three-step reaction involving arginine decarboxylase (ADC) (Hashimoto et al. 1989), agmatine deiminase (AIH), and N-carbamoyl putrescine amidase (CPA) (Kohnen-Johannsen and Kayser 2019). The next step in TA biosynthesis involves the enzyme putrescine methyltransferase (PMT), which catalyzes the methylation of putrescine to produce N-methylputrescine. N-methylputrescine oxidase (MPO) then converts N-methylputrescine into 4-methylaminobutanal (Katoh et al. 2007). Spontaneous cyclization of 4-methylaminobutanal is thought to have resulted in the formation of N-methyl-Δ^1-pyrrolinium cation (Bedewitz et al. 2014 and references therein). The activated pyrrolinium moiety undergoes two rounds of malonyl-coenzyme A-mediated condensation that is catalyzed by a type III polyketide synthase

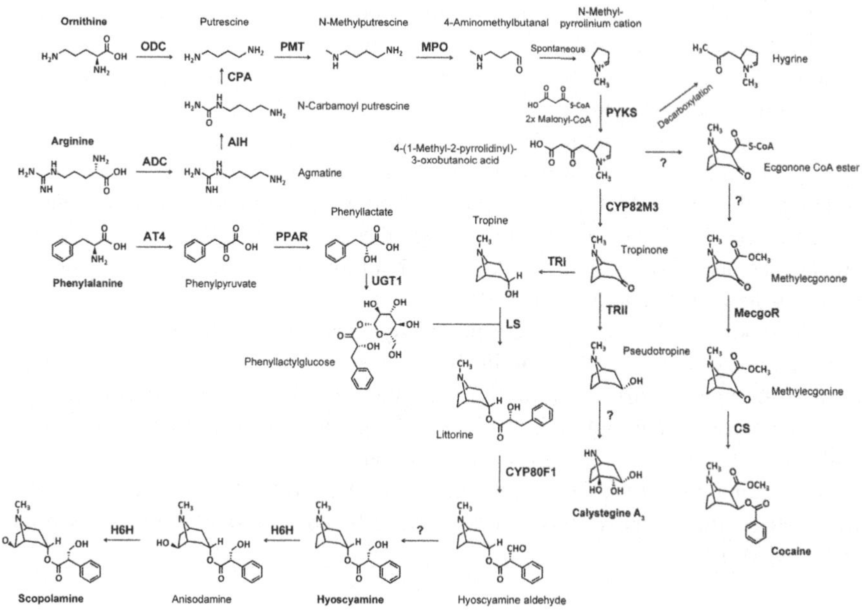

Fig. 8.1 Biosynthetic pathways of hyoscyamine, scopolamine, calystegines, and cocaine in TA-producing plants. *ODC* ornithine decarboxylase, *ADC* arginine decarboxylase, *AIH* agmatine deiminase, *CPA* N-carbamoyl putrescine amidase, *PMT* putrescine N-methyltransferase, *MPO* N-methylputrescine oxidase; *PYKS* type III polyketide synthase; *CYP82M3* tropinone synthase, *TR-I* tropinone reductase I, *TR-II* tropinone reductase II, *AT4* aromatic amino acid aminotransferase 4, *PPAR* phenylpyruvic acid reductase, *UGT1* phenyllactate UDP-glycosyltransferase, *LS* littorine synthase, *CYP80F1* littorine mutase, *H6H* hyoscyamine 6b-hydroxylase, *MecgoR* methylecgonone reductase, *CS* cocaine synthase. Uncharacterized TA biosynthesis enzymes are represented by "?"

(PYKS) to form 4-(1-methyl-2-pyrrolidinyl)-3-oxobutanoic acid, which is converted to tropinone via a cytochrome P450-mediated catalysis by CYP82M3 (Bedewitz et al. 2018). The decarboxylation of the oxobutanoic acid intermediate resulted in the production of hygrine (Fig. 8.1), while the formation of cuscohygrine (not shown) is likely through decarboxylative condensation of the oxobutanoic acid intermediated with a second N-methyl-Δ^1-pyrrolinium cation (Bedewitz et al. 2018). The formation of tropinone is a key branch point in the TA biosynthetic pathway (Bedewitz et al. 2014). Tropinone is reduced by tropinone reductase I (TR-I) to produce tropine, while tropinone reductase II (TR-II) reduces tropinone to pseudotropine (Bedewitz et al. 2014). Littorine synthase (LS) catalyzed the condensation of tropine with phenyllactylglucose to form littorine (Qiu et al. 2020). Phenyllactylglucose is derived from the amino acid phenylalanine, which undergoes transamination to yield phenylpyruvate, a reaction catalyzed by aromatic amino acid aminotransferase (AT4) (Bedewitz et al. 2014). Phenylpyruvic acid reductase (PPAR) catalyzes the reduction of phenylpyruvate to phenyllactate (Qiu et al. 2018), which then undergoes glycosylation with the help of UDP-glycosyltransferase (UGT1) to form phenyllactylglucose (Qiu et al. 2020). The intermediate littorine then undergoes P450-mediated rearrangement to produce hyoscyamine aldehyde, which is converted to hyoscyamine via uncharacterized alcohol dehydrogenase. Hyoscyamine is converted to scopolamine via a two-step epoxidation reaction catalyzed by hyoscyamine 6β-hydroxylase (H6H) (Fig. 8.1). In *Cochlearia*, cochlearine is produced instead of hyoscyamine (Brock et al. 2006). The biosynthesis of hyoscyamine and scopolamine in the Solanaceae family is now almost completely elucidated (Zhao et al. 2020). Unlike hyoscyamine and scopolamine—which are derived from tropine—calystegines are directly synthesized from pseudotropine (Dräger 2004). Beyond this, nothing is known about the biosynthesis of calystegines (Kohnen-Johannsen and Kayser 2019).

The hypothetical pathway of cocaine biosynthesis follows the same initial steps seen in hyoscyamine and scopolamine biosynthesis, with either ornithine or arginine as the starting precursor up to the formation of 4-(l-methyl-2-pyrrolidinyl)-3-oxobutanoic acid (Schmidt et al. 2015) (Fig. 8.1). The oxobutanoic acid intermediate undergoes cyclization to form methylecgonone, which is then reduced to methylecgonine by methylecgonone reductase (MecgoR) (Jirschitzka et al. 2012). Methylecgonine is finally converted to cocaine by cocaine synthase (CS) (Schmidt et al. 2015) (Fig. 8.1).

8.4 Tropane Alkaloid Biosynthesis Pathway Genes and their Regulation

8.4.1 Genes in the Early Steps of TA Biosynthesis

8.4.1.1 Ornithine decarboxylase (ODC)

All TAs originate from ornithine and/or arginine through common early biosynthetic pathways, mediated by ODC and ADC (Fig. 8.1). These early pathways resulted in the ubiquitous symmetric diamine putrescine, which is readily metabolized to polyamines, spermidine, and spermine (Hashimoto and Yamada 1994). ODC occurs in all living organisms (Facchini 2001). *ODC* genes were reported from several plant species including TA-producing plants, viz., *A. belladonna*, *H. niger*, *D. stramonium*, and *E. coca* (Michael et al. 1996; Docimo et al. 2012; Zhao et al. 2019, 2020). The first plant ODC cDNA was cloned from *D. stramonium* (Michael et al. 1996). The predicted DsODC was similar to other ODCs and ADCs as well. *DsODC* exhibited higher expression in rapidly growing root compared to stem or leaf. Similarly, in *H. niger*, *HnODC* was expressed predominately in root and upregulated by methyl jasmonate (MeJA) treatment (Zhao et al. 2019). MeJA application enhanced the synthesis of putrescine and *N*-methylputrescine in *H. niger*; however, hyoscyamine and scopolamine level remains unaltered (Zhao et al. 2019). Interestingly, in *E. coca*, *EcODC* transcripts were highest in young leaf, followed by bud and mature leaf. *EcODC* expression was negligible in stem and root (Docimo et al. 2012). Though two routes (arginine or ornithine pathway) to putrescine biosynthesis exist (Facchini 2001), the preferential route to putrescine biosynthesis has been unclear (Chintapakorn and Hamill 2007). In *D. stramonium*, ADC enzyme was reported to be more important for TA biosynthesis (Robins et al. 1991). In contrast, more recently, enzyme inhibitor-based feeding assay suggested that ODC, rather than ADC, participates in TA biosynthesis in *A. belladonna* (Zhao et al. 2020). Moreover, silencing or overexpressing *AbODC* in *A. belladonna* root culture resulted in either reduction or increased production of putrescine, and hyoscyamine, respectively (Zhao et al. 2020).

8.4.1.2 Arginine Decarboxylase (ADC), Agmatine Deiminase (AIH), and N-Carbamoyl Putrescine Amidase (CPA)

To form putrescine, the amino acid arginine undergoes three-step reactions, including decarboxylation by ADC to form agmatine which is hydrolyzed sequentially by AIH and CPA to form putrescine (Fig. 8.1). ADC is absent in mammals and many lower eukaryotes (Facchini 2001). The first *ADC* gene was cloned from *Escherichia coli* (Moore and Boyle 1990). *ADC* cDNAs have been isolated from TA-yielding plants such as *E. coca* (Docimo et al. 2012) and *H. niger* (Zhao et al. 2019). Numerous studies linked *ADC* in the plant responses to abiotic stresses (Paschalidis et al. 2019). *Hyoscyamus niger* has two *ADC* genes (*HnADC1* and *HnADC2*). The *HnADC1* and *HnADC2* transcripts were found in both root and leaf with no significant difference. Also, *HnADC1* and *HnADC2* expressions were unchanged by MeJA (Zhao et al. 2019). In contrast, the expression of *E. coca ADC* was higher in the bud and young leaf, compared to stem and root. The expression was least in mature leaf (Docimo et al. 2012). *EcADC* was fished out from an *E. coca* young leaf cDNA library using radiolabelled *Nicotiana tabacum ODC* as the probe. Several transcripts for *AIH* and *CPA* were identified in the root transcriptome of *Anisodus acutangulus* (Cui et al. 2015) and *A. belladonna* (Bedewitz et al. 2014).

8.4.1.3 Putrescine Methyltransferase (PMT)

Putrescine methyltransferase catalyzes the *N*-methylputrescine conversion to putrescine, which is the first committed step in the TA biosynthetic pathway (Fig. 8.1). PMT enzyme activity was detected only in plants (Biastoff et al. 2009). The first *PMT* gene was cloned from *N. tabacum* (Hibi et al. 1994). Similar to many genes involved in secondary metabolism, *PMT* genes were predominantly expressed in specific organ and tissues (Suzuki et al. 1999a; Biastoff et al. 2009; Zhao et al. 2017; Geng et al. 2018); however, an exceptional constitutive expression was reported in *A. acutangulus* (Kai et al. 2009). Restricted expression of *PMT* genes to plant roots were shown for *A. belladonna* (Suzuki et al. 1999a), *D. stramonium* (Biastoff et al. 2009), *H. niger* (Geng et al. 2018), and *Scopolia lurida* (Zhao et al. 2017). Kai et al. (2009) reported constitutive expression of *AaPMT* in leaf, stem, and root of *A. acutangulus*. In *D. myoporoides*, *DmPMT* transcripts were higher in root compared with leaf and stem (Kohnen et al. 2018). In *A. belladonna* and *H. niger*, *PMT* gene expression was induced by MeJA (Suzuki et al. 1999a; Geng et al. 2018). Attempts to alter the metabolic pathway by increasing PMT enzyme activity alone were unsuccessful. Transgenic *A. belladonna* root culture with high PMT activity exhibits enhanced N-methylputrescine content; however, the alkaloid content and profile were the same as the control (Sato et al. 2001; Rothe et al. 2003). Similar results were observed in transgenic *Duboisia* hybrid root culture overexpressing *PMT* (Moyano et al. 2002). But interestingly, when *PMT* and *hyoscyamine-6-β-hydroxylase* (*H6H*) were simultaneously introduced and overexpressed, a significantly higher accumulation of scopolamine was observed in *A. belladonna* (Xia et al. 2016) and *H. niger* (Zhang et al. 2004) transgenic.

8.4.1.4 N-Methylputrescine Oxidase (MPO)

MPO genes were first characterized from the species *N. tabacum* (Heim et al. 2007; Katoh et al. 2007). *NtMPO* genes belong to a small multigene family. Tobacco *MPO* genes were expressed specifically in the root and induced by MeJA (Heim et al. 2007; Katoh et al. 2007). Although MPO enzymes have been partially purified from the roots of *H. niger* and *Brugmansia candida* x *aurea* hybrid (Hashimoto et al. 1990, Boswell et al. 1999), *MPO*-like genes are yet to be discovered.

8.4.2 Genes Involved in Hyoscyamine and Scopolamine Biosynthesis

8.4.2.1 Polyketide Synthase (PYKS) and Cytochrome P450 Enzyme (CYP82M3)

More recently, the missing pathways and enzymes linking N-methyl-Δ^1-pyrrolinium to the synthesis of tropinone were revealed. Tropinone is the first key tropane intermediate in the biosynthesis of hyoscyamine and scopolamine (Hashimoto and Yamada 1994). Bedewitz et al. (2018) showed that a type III polyketide synthase (AbPYKS) and a cytochrome P450 (AbCYP82M3) simultaneously catalyzed the N-methyl-Δ^1-pyrrolinium conversion to tropinone. *AbPYKS* and *AbCYP82M3* were obtained through mining of an *A. belladonna* root transcriptome assembly reported previously by Bedewitz et al. (2014). *AbPYKS* and *AbCYP82M3* showed significant preferential expression in lateral roots relative to leaf and stem. The root-preferential expression was validated by real-time PCR. *AbPYKS* predicts for a protein with high amino acid sequence similarity to *Solanum lycopersicum* chalcone synthase B; the function of chalcone synthase B is unknown. *AbCYP82M3* encodes CYP82 subfamily members that are related to NtCYP82E4 (a nicotine demethylases) from *N. tabacum* (Bedewitz et al. 2018). Huang et al. (2019) reported the isolation of three PYKSs, namely, *A. acutangulus* PYKS, *D. stramonium* PYKS, and *A. belladonna* PYKS, and obtained the crystal structure of AaPYKS protein.

8.4.2.2 Tropinone Reductase (TR)

The two stereospecific tropinone reductases (TR-I and TR-II) constitute a branch point in TA metabolism (Fig. 8.1). TRs belong to short chain dehydrogenase/reductase enzyme family that catalyzes NAD(P)H-dependent oxidoreductase reactions (Dräger 2006). Tropinone serves as substrate for both TR-I and TR-II (Hashimoto et al. 1992). TR-I catalyzes the reduction of tropinone to tropine, whereas TR-II forms pseudotropine. Tropine is used to generate scopolamine, while pseudotropine is the precursor of calystegine biosynthesis. A gene duplication event resulted in these two different TRs in Solanaceae (Nakajima et al. 1993). *D. stramonium* TR-I and TR-II are 273 and 260 amino acids long, respectively, and are 64% identical (Nakajima et al. 1993). Southern hybridization showed that *TR* genes were absent in non-TA-producing Solanaceae species like *N. tabacum* (Nakajima et al. 1993). Recently, cDNAs coding for TRs were also isolated from several TA-producing plants including *A. acutangulus* (Kai et al. 2009), *B. arborea* (Qiang et al. 2016), *D. myoporoides* (Kohnen et al. 2018), *Przewalskia tangutica* (Wu et al. 2019) and *S. lurida* (Zhao et al. 2017). Hashimoto et al. (1992) showed high TR-I and TR-II enzyme activities in the root of *A. belladonna*, *D. stramonium*, and *H. niger*. TR-I enzyme activity was strongest in the young root, whereas high TR-II enzyme activity was found in both root and stem of *D. stramonium* and *A. belladonna*. TR-II enzyme activity was higher than TR-I enzyme activity in *H. niger* and *A. belladonna*, whereas in *D. stramonium* the reverse relation was found. Similarly, in several Solanaceae plants, *TRs* exhibited highest transcript abundance in roots as compared to stem and leaf (Kai et al. 2009; Qiang et al. 2016; Zhao et al. 2017; Kohnen et al. 2018; Wu et al. 2019). Additionally, immunoblot analyses revealed that the two TRs showed cell-specific localization in *H. niger* root (Nakajima and Hashimoto 1999). HnTR-I was found in the outer cell layers of the cortex and endodermis. HnTR-II, however, showed preferentially localization in the inner cortex layers, endodermis, and pericycle.

Transgenic *A. belladonna* root cultures overexpressing *DsTR-I* alter the alkaloid composition (Richter et al. 2005). Transgenic root contained threefold and fivefold more hyoscyamine and scopolamine, respectively, while calystegine content was lesser by 30–90% of untransformed control root. Data suggested that transformation with TR-I gene could potentially alter the ratio of TAs versus pseudotropine. Similarly, overexpressing *SlTR-I* significantly increases the hyoscyamine and scopolamine levels in *S. lurida* root cultures (Zhao et al. 2017). Kai et al. (2012) reported the co-introduction of genes coding for AaTR-I and the downstream rate-limiting enzyme AaH6H into *A. acutangulus* hairy roots. Overexpressing both *AaTR-1* and *AaH6H* resulted in a 4.49-fold higher accumulation of TAs compared with the untransformed control. This finding showed that transformation with a combination of key rate-controlling genes could produce desirable synergistic effect stimulating TA biosynthesis.

8.4.2.3 Littorine Synthase (LS), Aromatic Amino Acid Aminotransferase 4 (AT4), Phenylpyruvic Acid Reductase (PPAR), and Phenyllactate UDP-Glycosyltransferase (UGT1)

Formation of a key intermediate, littorine from phenyllactate and tropine, was completely elucidated, only recently (Fig. 8.1). Qiu et al. (2020) reported the identification and functional characterization of two novel genes that encode for a UGT1 and LS. UGT1, which catalyzes phenyllactate and UDP-glucose ligation to produce phenyllactylglucose and LS, then condenses phenyllactylglucose and tropine through esterification yielding littorine (Fig. 8.1). *AbUGT1* and *AbLS* genes were initially identified using the hidden Markov model (HMM) search program against the publically available *A. belladonna* transcriptome data available at "http://medicinalplantgenomics.msu.edu/index.shtml." HMM search was performed using the domain profiles for UDP-glycosyltransferase (PF00201) and serine carboxypeptidase (SCP)-like acyltransferases (PF00450). *AbUGT1* and *AbLS* transcripts were abundant specifically in secondary root tissues. Suppression of either *AbUGT1* or *AbLS* through VIGS and RNAi disrupted the biosynthesis of littorine, hyoscyamine, and scopolamine. Recombinant AbUGT1 exhibits in vitro glycosylation of phenyllactate. Co-expression *of AbUGT1* and *AbLS* gene in the TA-lacking plant *N. benthamiana* led to littorine accumulation when both phenyllactate and tropine were supplied (Qiu et al. 2020).

Precursor phenyllactate is derived from amino acid phenylalanine via two-step reactions (Fig. 8.1). First phenylalanine is transaminated to phenylpyruvate, the reaction is catalyzed by AbAT4 (Bedewitz et al. 2014), and PPAR then catalyzes the reduction of phenylpyruvate to phenyllactate (Qiu et al. 2018). *AbAT4* and *AbPPAR* genes were obtained through the mining of the *A. belladonna* root transcriptome assembly (Bedewitz et al. 2014; Qiu et al. 2018). *AbAT4* showed higher transcript level in roots and sterile seedlings compared to non-root tissue (Bedewitz et al. 2014). *AbPPAR* transcripts were specifically found in root endodermis and pericycles (Qiu et al. 2018). Silencing of *AbAT4* and *AbPPAR* resulted in the disruption of hyoscyamine, scopolamine, and/anisodamine synthesis through the reduction of phenyllactate levels. Additionally, purified recombinant AbAT4 catalyzed the transamination of phenylalanine to phenylpyruvate in vitro, while purified His-tagged AbPPAR was able to catalyze the reduction of phenylpyruvate to phenyllactate in a substrate-feeding assay (Bedewitz et al. 2014; Qiu et al. 2018).

8.4.2.4 Littorine Mutase/Monooxygenase/CYP80F1

In 2006, Li et al. demonstrated that CYP80F1 (also known as littorine mutase or monooxygenase) catalyzes littorine oxidation to hyoscyamine aldehyde. They hypothesize that an additional second enzyme, most probably an alcohol dehydrogenase, is required to convert hyoscyamine aldehyde to hyoscyamine (Fig. 8.1). *HnCYP80F1* was identified through functional genomics approach involving VIGS. The root-specific partial *HnCYP80F1* was isolated from a subtracted cDNA library enriched for *H. niger* root transcriptome. Using the partial *HnCYP80F1* sequence information, 5′ and 3′ RACE and PCR amplification were performed to obtain the full-length *HnCYP80F1* cDNA. The predicted HnCYP80F1 shares sequence similarity to *Coptis japonica* (S)-N-methylcoclaurine-3′-hydroxylases. The suppression of the *HnCYP80F1* transcripts by VIGS and RNAi led to littorine accumulation and reduction of hyoscyamine levels (Li et al. 2006).

8.4.2.5 Hyoscyamine 6β-Hydroxylase (H6H)

Hyoscyamine 6β-hydroxylase (H6H) converted hyoscyamine into the epoxide scopolamine (Fig. 8.1). H6H is a 2-oxoglutarate-dependent dioxygenase, first purified from *H. niger*, a scopolamine-rich plant (Hashimoto and Yamada 1986; Hashimoto et al. 1993). This enzyme was localized in the root pericycle (Hashimoto et al. 1991). *HnH6H* transcripts were highly abundant in the root but absent in stem and leaf (Matsuda et al. 1991). *A. belladonna H6H* expression was detected in root and anther; however, in stem, leaf, and other floral parts, the gene expression was absent (Suzuki et al. 1999b). *S. lurida H6H* exhibits exclusive expression in the root (Zhao et al. 2017). In contrast, the *D. myoporoides H6H* transcripts were most abundant in leaves of mature plant (Kohnen et al. 2018). This may explain the observed accumulation of scopolamine in the leaves of *D. myoporoides* (Griffin and Lin 2000; Foley 2006). Kanegae et al. (1994) also reported the species-dependent expression of *H6H*. Overexpressing *H6H* resulted in improving the scopolamine biosynthesis and TA production in *A. belladonna* and *Duboisia* hybrid (Palazón et al. 2003; Xia et al. 2016).

8.4.3 Genes Involved in Cocaine Biosynthesis

Cocaine biosynthesis pathway, from its branch point with the intermediate substrate N-methyl-Δ^1-pyrrolinium, is still not completely elucidated (Fig. 8.1). Only the genes involved in the final step and the penultimate step in cocaine biosynthesis are well characterized (Jirschitzka et al. 2012; Schmidt et al. 2015).

8.4.3.1 Methylecgonone Reductase (MecgoR)

Methylecgonone reductase (*MecgoR*) gene encodes for a protein of 327-amino-acid residues (Jirschitzka et al. 2012). The predicted protein is similar to reported aldo-keto reductases including chalcone reductase and codeinone reductase enzymes of flavonoid and morphine alkaloid biosynthesis, respectively. All plant aldo-keto reductases have α/β-barrel motif and a cofactor NADH or NADPH (Jez et al. 1997). MecgoR reduces methylecgonone stereospecifically to 2-methylecgonine (Fig. 8.1). It exhibits the highest protein activity and gene expression level in the young, growing leaf of *E. coca*. The MecgoR activity and *MecgoR* expression were not detected in the root, which is the site of TA metabolism in the Solanaceae (Jirschitzka et al. 2012; Kohnen-Johannsen and Kayser 2019). The MecgoR protein is found restricted to the spongy mesophyll and palisade tissue of young, growing leaf and sepal (Jirschitzka et al. 2012).

8.4.3.2 Cocaine Synthase (CS)

Cocaine synthase (CS) catalyzed the esterification of methylecgonine with benzoic acid to form cocaine (Fig. 8.1). CS is a BAHD acyltransferase (D'Auria 2006) and can produce cocaine by activated benzoyl-CoA thioester and cinnamoylcocaine through activated cinnamoyl-CoA thioesters (Schmidt et al. 2015). *CS* was identified from *E. coca* young leaf transcriptome generated through 454 sequencing; *CS* exhibits the highest gene expression level in young, growing leaf than in mature leaf, stem, and flower; and expression was negligible in the root (Schmidt et al. 2015). CS enzyme activity was specifically localized to the palisade parenchyma and spongy mesophyll of young, growing leaf (Schmidt et al. 2015). These tissues were the sites for the biosynthesis and storing of cocaine and other TAs in *E. coca* (Torre et al. 2013).

8.4.4 Calystegine Biosynthesis Genes

The detailed biosynthesis of calystegine is not known (Fig. 8.1). Tropinone is reduced to precursor pseudotropine, a reaction catalyzed by TR-II (Nakajima et al. 1993; Hashimoto and Yamada 1994; Scholl et al. 2001). A TR-II gene has been isolated and characterized from *C. sepium*, which showed an exclusive preference for pseudotropine formation (Dräger 2004). Other TR-like sequences have also been isolated from *C. sepium*, one of which reduces tropinone to tropine, indicating that both TR-I and TR-II genes are present in *C. sepium* (Dräger 2004). Further attempts to decode the biosynthetic pathway of these compounds have not been reported so far (Kohnen-Johannsen and Kayser 2019). This may be due to lesser medical use compared to pharmacologically important scopolamine or cocaine (Scholl et al. 2001; Kohnen-Johannsen and Kayser 2019).

8.5 Identification and Elucidation of the Pathway Genes through Transcriptome Analysis

Although TAs are produced by several species of plants from across many families (Table 8.1), the most prominent TAs are found mainly in species belonging to the Erythroxylaceae and Solanaceae families (Fodor and Dharanipragada 1990; Griffin and Lin 2000). Therefore, past investigations on TA biosynthetic pathways and pathway genes had focused mostly on a selected small group of plant species. In recent years, transcriptome analyses of few selected TA-rich species including *A. belladonna*, *H. niger*, *A. acutangulus*, *D. stramonium*, and *E. coca* have aided in the identification of key novel biosynthetic genes that play crucial roles in different stages of TA metabolism. For instance, mining of the *A. belladonna* transcriptome assembly helped identify the unique *AbAT4* gene. AbAT4 is the crucial enzyme channeling the amino acid phenylalanine to the formation of littorine, a key intermediate in the scopolamine biosynthesis (Bedewitz et al. 2014). The root-specific co-expression of *AbAT4* with other previously known TA pathway genes aided in its discovery. For this work, a total of 13 transcriptome corresponding to 11 tissue types including leaf, stem, flower, root, and fruit of mixed developmental stages of *A. belladonna* were generated through RNA-seq on the Illumina platform. De novo assembly generated 80,624 reliable transcripts that represented 43,951 unigenes. The *A. belladonna* transcriptome datasets, including the assemblies and transcript abundance values, were made open access at http://medicinalplantgenomics.msu.edu/ (Bedewitz et al. 2014). More recently, unigenes encoding for a type III AbPYKS and cytochrome P450 (AbCYP82M3) which catalyze the formation of the intermediate tropinone (Fig. 8.1) in the TA biosynthesis pathway were identified from the root transcriptome of the abovementioned transcriptome data assembly (Bedewitz et al. 2018). Interestingly, using the same *A. belladonna* root transcriptome data publically available at http://medicinalplantgenomics.msu.edu/, several novel genes of the hyoscyamine and scopolamine biosynthetic pathways were isolated and characterized, namely, *AbPPAR* (Qiu et al. 2018), *AbUGT1* (Qiu et al. 2020), and *AbLS* (Qiu et al. 2020). In a recent study, three PYKSs were isolated from RNA-seq-generated hairy root transcriptomes of three different hyoscyamine- and scopolamine-producing plants, namely, *A. acutangulus*, *D. stramonium*, and *A. belladonna* (Huang et al. 2019). Based on crystal structure and biochemical activity of recombinant AaPYKS, the detailed mechanistic model for the condensation of malonyl-CoA with N-methyl-Δ^1-pyrrolinium during the formation of intermediate 4-(1-methyl-2-pyrrolidinyl)-3-oxobutanoic acid in the TA biosynthetic pathway was elucidated (Huang et al. 2019). In a couple of studies using transcriptome analyses, *HnPMT* (Geng et al. 2018), *HnODC* (Zhao et al. 2019), and a novel *HnCYP80F1* (*littorine mutase*) involved in TA biosynthesis were isolated from root tissues of *H. niger* (Li et al. 2006). Littorine mutase catalyzes the formation of hyoscyamine aldehyde from littorine (Fig. 8.1).

Unlike in members of Solanaceae where TAs are produced in roots, the biosynthesis of TAs in Erythroxylaceae members occurs in the aerial tissues. The evidence for this distinction was consolidated by the identification of a cocaine biosynthesis enzyme (CS, cocaine synthase) through screening of *E. coca* young leaf transcriptome (Schmidt et al. 2015).

To further understand TA biosynthesis mechanism, Cui et al. (2015) performed deep RNA sequencing of *A. acutangulus* root tissue. *A. acutangulus* root was reported to contain high concentration of total TAs equivalent to 1.2% of the dry weight (Kai et al. 2012). The study yielded 90,903 unigenes having average length of 592 base pairs. After functional annotation and pathway analysis, 48 unigenes were categorized to tropane, pyridine, and piperidine alkaloid biosynthesis, and 82 unigenes were cytochrome P450 enzyme involved in secondary metabolism. Additionally, over 1000 unigenes encoding transcriptional factors belonging to AP2/ERF, bHLH, MYB, and WRKY families implicated in secondary metabolite biosynthesis were identified. Some of these transcriptional factors might have roles as master regulator to fine-tune TA biosynthesis to developmental and environmental cues. *A. acutangulus* orthologs of *CYP80F1*, *diamine oxidase*, *alcohol dehydrogenase*, and *AT4* were also identified from the root transcriptome (Cui et al. 2015).

The transcriptomic information from less studied plant species can also be potentially used for determining the presence or absence of biosynthetic pathways for TAs and other bioactive secondary compounds. Bioinformatics analyses of transcriptome datasets from *Corydalis yanhusuo* (Papaveraceae), a medicinal plant, helped detect unigenes for several alkaloid biosynthesis pathways, including 30 unigenes for TA biosynthesis (Liao et al. 2016). Comparative transcriptomic studies of diverse TA-producing plants could provide deeper insights into the biogenesis pathways for other lesser studied TAs.

8.6 Promoter Analysis of TA Biosynthesis Genes

As mentioned earlier, biosynthesis of TAs and other secondary metabolites is usually localized in specific tissues (Ziegler and Facchini 2008; Kohnen-Johannsen and Kayser 2019). Analysis of promoter activities of key biosynthetic genes can provide clear evidence of the sites of TA biosynthesis within the plants. Additionally, gene promoters can be used for modulating the overall biosynthetic pathways through biotechnological interventions. A few promoters of the TA biosynthetic pathway genes have aided in elucidating the gene function and in determining the localization of the secondary metabolite biosynthesis in the studied plants. The *A. belladonna H6H* gene promoter::GUS transgene analysis further showed specific expression in pollen mother cells, tapetum, and root pericycle (Suzuki et al. 1999b). Similarly, *H. niger* PMT1 promoter::GUS transgene exhibits specific expression in the root of transgenic *Arabidopsis thaliana* and *H. niger* (Geng et al. 2018). Concomitant with the tissue-specific expression, the promoter regions of *AbH6H* and *HnPMT1* have several putative *cis*-acting regulatory motifs involved in auxin, MeJA, and salicylic acid responsiveness and the activation of plant defense and stress genes such as the MYB binding site, TC-rich repeats, W-box, heat shock element, low-temperature response element, and tissue-specific elements (Suzuki et al. 1999b; Geng et al. 2018; Cui et al. 2015). Function prediction of corresponding transcriptional factors would facilitate in-depth understanding of the regulatory network of TA biosynthesis. Known promoters with tissue-specific activity can be potentially used for genetically modulating tissue-specific biosynthesis of TAs in specialized plant tissues (Jindal et al. 2015). With the increase in the number of elucidated genes and our understanding of the TA metabolism pathways in plants, the 5′ upstream promoter sequences of key genes encoding enzymes catalyzing the rate-limiting step could also provide useful prospects for targeted or systemic metabolic engineering to enhance TA production in plants.

8.7 Conclusion

The generation and analyses of transcriptomic data have facilitated significant advances in plant functional genomics studies. In many plant species, the availability of transcriptome data allows for the functional annotation of assembled transcripts and helps to identify potential pathway genes and their molecular functions in different biological processes (Bedewitz et al. 2014; Paul et al. 2014). Transcriptomic information is particularly important for understanding the role of key biosynthetic genes involved in secondary metabolism. However, many secondary metabolites are tissue-specific which prompts for analyses of tissue-specific transcriptomes (Bedewitz et al. 2014; Cui et al. 2015; Kohnen et al. 2018). Secondary metabolites are products of an innate plant defense response mechanism, and analyses of transcriptomes posttreatment with exposure to specific stressors become necessary to capture the elicited transcript/s. Additionally, a single type of secondary metabolites, such as TAs, may be produced by different species of plants albeit at varying levels. Intriguingly, new studies in metabolite correlation networks increasingly show that the TA biogenesis is intricately complex and finely controlled (Nguyen et al. 2015). This necessitates expanded screening of more transcriptomes for hitherto unknown genes potentially involved in the intricate biosynthetic pathways by employing powerful functional genomics tools, such as VIGS-based screening (Senthil-Kumar and Mysore 2011). It also becomes pertinent to merge transcriptomic and metabolomics by utilizing efficient techniques to generate more tissue-specific transcriptomes, such as hairy roots (Bedewitz et al. 2014; Geng et al. 2018), and robust sample harvesting methods for metabolite profiling, such as vascular sap collection (Longchar et al. 2020), to study the spatiotemporal translocation of metabolite intermediates observed in *D. myoporoides* (Kohnen et al. 2018) and *P. tangutica* (Wu et al. 2019). In addition to transgenic plants overexpressing key TA biosynthetic genes of interest (Sato et al. 2001; Zhang et al. 2004; Kai et al. 2012; Zhao et al. 2020), current approaches to TA metabolic engineering can also now target heterologous expression of the pathway gene(s) cassette in *E. coli* or yeast to produce most valuable scopolamine outside of the plant (Cardillo et al. 2017). In recent years, mining of the transcriptome data of TA-producing plants such as *A. belladonna* and *H. niger* has led to a near-complete elucidation of the biosynthesis of hyoscyamine and scopolamine. Transcriptome analysis of diverse TA-producing plants would shed light on occurrence TA in non-Solanaceae family (Table 8.1) such as in Brassicaceae, Convolvulaceae, Moraceae, Proteaceae, and Rhizophoraceae. It is further envisaged that more increasing transcriptome studies would also facilitate the eventual deciphering of calystegine and cocaine biosynthesis pathways.

Acknowledgments AP duly acknowledges the facilities provided by the Department of Botany, Nagaland University, Lumami. BL is grateful to the Principal, Pachhunga University College, and the Vice-Chancellor, Mizoram University, Aizawl, for providing institutional support to write this article. JD acknowledges Dr. Sanjay Kumar, Director, CSIR-IHBT, Palampur, for providing the necessary facilities needed during the preparation of this book chapter. This book chapter represents CSIR-IHBT communication number 4630.

References

Al-Said MS, Evans WC, Grout RJ (1989) Alkaloids of *Erythroxylum hypericifolium* stem bark. Phytochemistry 28:671–673

Arbain D, Wiryani RD, Sargent MV (1991) A new tropane alkaloid from *Pellacalyx axillaris*. Aust J Chem 44:1013–1015

Asano N, Oseki K, Tomioka E, Kizu H, Matsui K (1994) N-containing sugars from *Morus alba* and their glycosidase inhibitory activities. Carbohydr Res 259:243–255

Backlund A (2010) Topical chemical space in relation to biological space. In: Mander L, Lui HW (eds) Comprehensive natural products II: chemistry and biology, vol 3. Elsevier, Oxford, pp 47–79

Bedewitz MA, Góngora-Castillo E, Uebler JB, Gonzales-Vigil E, Wiegert-Rininger KE, Childs KL, Hamilton JP, Vaillancourt B, Yeo YS, Chappell J, DellaPenna D, Jones AD, Buell CR, Barry CS (2014) A root-expressed L-phenylalanine:4-hydroxyphenylpyruvate aminotransferase is required for tropane alkaloid biosynthesis in *Atropa belladonna*. Plant Cell 26:3745–3762

Bedewitz MA, Jones AD, D'Auria JC, Barry CS (2018) Tropinone synthesis via an atypical polyketide synthase and P450-mediated cyclization. Nat Commun 9:5281

Biastoff S, Brandt W, Dräger B (2009) Putrescine N-methyltransferase-the start for alkaloids. Phytochemistry 70:1708–1718

Bick IRC, Gillard JW, Leow HM, Preston NW (1979) Alkaloids of *Agastachys odorata*. Aust J Chem 32:2071–2076

Boswell HD, Dräger B, McLauchlan WR, Portsteffen A, Robins DJ, Robins RJ, Walton NJ (1999) Specificities of the enzymes of N-alkyltropane biosynthesis in *Brugmansia* and *Datura*. Phytochemistry 52:871–878

Brock A, Herzfeld T, Paschke R, Koch M, Dräger B (2006) Brassicaceae contain nortropane alkaloids. Phytochemistry 67:2050–2057

Cannon JR, Joshi KR, Meehan GV, Williams JR (1969) The tropane alkaloids from three Western Australian *Anthocercis* species. Aust J Chem 22:221–227

Cardillo AB, Perassoloa M, Sartuquia M, Taloua JR, Giulietti AM (2017) Production of tropane alkaloids by biotransformation using recombinant *Escherichia coli* whole cells. Biochem Eng J 125:180–189

Castillo G, Cruz LL, Tapia-López R, Olmedo-Vicente E, Carmona D, Anaya-Lang AL, Fornoni J, Andraca-Gómez G, Valverde PL, Núñez-Farfán J (2014) Selection mosaic exerted by specialist and generalist herbivores on chemical and physical defense of *Datura stramonium*. PLoS One 9: e102478

Chang J, Xie W, Wang L, Ma N, Cheng S, Xie J (2006) An efficient approach to the asymmetric total synthesis of (−)-anisodine. Eur J Med Chem 41:397–400

Chintapakorn Y, Hamill JD (2007) Antisense-mediated reduction in ADC activity causes minor alterations in the alkaloid profile of cultured hairy roots and regenerated transgenic plants of *Nicotiana tabacum*. Phytochemistry 68:2465–2479

Cui L, Huang F, Zhang D, Lin Y, Liao P, Zong J, Kai G (2015) Transcriptome exploration for further understanding of the tropane alkaloids biosynthesis in *Anisodus acutangulus*. Mol Gen Genomics 290:1367–1377

D'Auria JC (2006) Acyltransferases in plants: a good time to be BAHD. Curr Opin Plant Biol 9:331–340

Docimo T, Reichelt M, Schneider B, Kai M, Kunert G, Gershenzon J, D'Auria JC (2012) The first step in the biosynthesis of cocaine in *Erythroxylum coca*: the characterization of arginine and ornithine decarboxylases. Plant Mol Biol 78:599–615

Dräger B (2004) Chemistry and biology of calystegines. Nat Prod Rep 21:211–223

Dräger B (2006) Tropinone reductases, enzymes at the branch point of tropane alkaloid metabolism. Phytochemistry 67:327–337

Dräger B, Funck C, Hoehler A, Mrachatz G, Nahrstedt A, Portsteffen A, Schaal A, Schmidt R (1994) Calystegines as a new group of tropane alkaloids in Solanaceae. Plant Cell Tissue Organ Cult 38:235–240

Eich E (2008) Solanaceae and convolvulaceae: secondary metabolites. Biosynthesis, chemotaxonomy, biological and economic significance. Springer, Berlin

El Imam YMA, Evans WC (1984) Tropane alkaloids of species of *Anthocercis*, *Cyphanthera* and *Crenidium*. Planta Med 50:86–87

Facchini PJ (2001) Alkaloid biosynthesis in plants: biochemistry, cell biology, molecular regulation, and metabolic engineering applications. Ann Rev Plant Physiol Plant Mol Biol 52:29–66

Fodor G, Dharanipragada R (1990) Tropane alkaloids. Nat Prod Rep 7:539–548

Foley P (2006) *Duboisia myoporoides*: the medical career of a native Australian plant. Hist Rec Aust Sci 17:31–69

Geng C, Zhao T, Yang C, Zhang Q, Bai F, Zeng J, Zhang F, Liu X, Lan X, Chen M, Liao Z (2018) Metabolic characterization of *Hyoscyamus niger* root-specific putrescine N-methyltransferase. Plant Physiol Biochem 127:47–54

Goldmann A, Milat ML, Ducrot PH, Lallamand JY, Maille M, Lepingle A, Charpin I, Tepfer D (1990) Tropane derivatives from *Calystegia sepium*. Phytochemistry 29:2125–2127

Griffin WJ, Lin GD (2000) Chemotaxonomy and geographical distribution of tropane alkaloids. Phytochemistry 53:623–637

Hanuš LO, Řezanka T, Spížek J, Dembitsky VM (2005) Substances isolated from Mandragora species. Phytochemistry 66:2408–2417

Hashimoto T, Yamada Y (1986) Hyoscyamine 6β-hydroxylase, a 2-oxoglutaratedependent dioxygenase, in alkaloid-producing root cultures. Plant Physiol 81:619–625

Hashimoto T, Yamada Y (1994) Alkaloid biogenesis: molecular aspects. Annu Rev Plant Physiol Plant Mol Biol 45:257–285

Hashimoto T, Yukimune Y, Yamada Y (1989) Putrescine and putrescine N-methyltransferase in the biosynthesis of tropane alkaloids in cultured roots of *Hyoscyamus albus*. I. Biochemical studies. Planta 178:123–130

Hashimoto T, Mitani A, Yamada Y (1990) Diamine oxidase from cultured roots of *Hyoscyamus niger*: its function in tropane alkaloid biosynthesis. Plant Physiol 93:216–221

Hashimoto T, Hayashi A, Amano Y, Kohno J, Iwanari H, Usuda S, Yamada Y (1991) Hyoscyamine 6β-hydroxylase, an enzyme involved in tropane alkaloid biosynthesis, is localized at the pericycle of the root. J Biol Chem 266:4648–4653

Hashimoto T, Nakajima K, Ongena G, Yasuyuki Y (1992) Two tropinone reductases with distinct stereospecificities from cultured roots of *Hyoscyamus niger*. Plant Physiol 100:836–845

Hashimoto T, Matsuda J, Yamada Y (1993) Two-step epoxidation of hyoscyamine to scopolamine is catalyzed by bifunctional hyoscyamine 6β-hydroxylase. FEBS Lett 329:35–39

Heim WG, Sykes KA, Hildreth SB, Sun J, Lu RH, Jelesko JG (2007) Cloning and characterization of a *Nicotiana tabacummethylputrescine oxidase* transcript. Phytochemistry 68:454–463

Hibi N, Higashiguchi S, Hashimoto T, Yamada Y (1994) Gene expression in tobacco low-nicotine mutants. Plant Cell 6:723–735

Huang JP, Fang C, Ma X, Wang L, Yang J, Luo J, Yan Y, Zhang Y, Huang SX (2019) Tropane alkaloids biosynthesis involves an unusual type III polyketide synthase and non-enzymatic condensation. Nat Commun 10:4036

Humam M, Kehrli T, Jeannerat D, Munoz O, Hosettmann K, Christen P (2011) Schizanthines N, O, and P, tropane alkaloids from the aerial parts of *Schizanthus tricolor*. J Nat Prod 74:50–53

Jenett-Siems K, Weigl R, Böhm A, Mann P, Tofern-Reblin B, Ott SC, Ghomian A, Kaloga M, Siems K, Witte L, Hilker M, Müller F, Eich E (2005) Chemotaxonomy of the pantropical genus *Merremia* (Convolvulaceae) based on the distribution of tropane alkaloids. Phytochemistry 66:1448–1464

Jez JM, Bennett MJ, Schlegel BP, Lewis M, Penning TM (1997) Comparative anatomy of the aldo-keto reductase superfamily. Biochem J 326:625–636

Jindal S, Longchar B, Singh A, Gupta V (2015) Promoters of AaGL2 and AaMIXTA-Like1 genes of *Artemisia annua* direct reporter gene expression in glandular and non-glandular trichomes. Plant Signal Behav 10:e1087629

Jirschitzka J, Schmidt GW, Reichelt M, Schneider B, Gershenzon J, D'Auria JC (2012) Plant tropane alkaloid biosynthesis evolved independently in the Solanaceae and Erythroxylaceae. Proc Natl Acad Sci U S A 109:10304–10309

Jirschitzka J, Dolke F, D'Auria JC (2013) Increasing the pace of new discoveries in tropane alkaloid biosynthesis. Adv Bot Res 68:39–72

Jocković N, Fischer W, Brandsch M, Brandt W, Dräger B (2013) Inhibition of human intestinal α-glucosidases by calystegines. J Agric Food Chem 61:5550–5557

Kai G, Zhang Y, Chen J, Li L, Yan X, Zhang R, Liao P, Lu X, Wang W, Zhou G (2009) Molecular characterization and expression analysis of two distinct putrescine N-methyltransferases from roots of *Anisodus acutangulus*. Physiol Plant 135:121–129

Kai G, Zhang A, Guo Y, Li L, Cui L, Luo X, Liu C, Xiao J (2012) Enhancing the production of tropane alkaloids in transgenic *Anisodus acutangulus* hairy root cultures by over-expressing tropinone reductase I and hyoscyamine-6β-hydroxylase. Mol Bio Syst 8:2883–2890

Kanegae T, Kajiya H, Amano Y, Hashimoto T, Yamada Y (1994) Species-dependent expression of the hyoscyamine 6 beta-hydroxylase gene in the pericycle. Plant Physiol 105:483–490

Katoh A, Shoji T, Hashimoto T (2007) Molecular cloning of *N-methylputrescine oxidase* from tobacco. Plant Cell Physiol 48:550–554

Kohnen KL, Sezgin S, Spiteller M, Hagels H, Kayser O (2018) Localization and organization of scopolamine biosynthesis in *Duboisia myoporoides* R. Br. Plant Cell Physiol 59:107–118

Kohnen-Johannsen KL, Kayser O (2019) Tropane alkaloids: chemistry, pharmacology, biosynthesis and production. Molecules 24:796

Li R, Reed DW, Liu E, Nowak J, Pelcher LE, Page JE, Covello PS (2006) Functional genomic analysis of alkaloid biosynthesis in *Hyoscyamus niger* reveals a cytochrome P450 involved in littorine rearrangement. Chem Biol 13:513–520

Liao D, Wang P, Jia C, Sun P, Qi J, Zhou L, Li X (2016) Identification and developmental expression profiling of putative alkaloid biosynthetic genes in *Corydalis yanhusuo* bulbs. Sci Rep 6:19460

Liebisch HW, Bernarsch H, Schuette HR (1973) Biosynthesis of tropane alkaloids. 12. Biosynthesis of cochlearine. Z Chemie 13:372–373

Longchar B, Phukan T, Yadav S, Senthil-Kumar M (2020) An efficient low-cost xylem sap isolation method for bacterial wilt assays in tomato. Appl Plant Sci 8:e11335

Lounasmaa M, Tamminem T (1993) The tropane alkaloids. In: Cordell GA (ed) The alkaloids: chemistry and pharmacology. Academic Press, San Diego, pp 1–115

Matsuda J, Okabe S, Hashimoto T, Yamada Y (1991) Molecular cloning of hyoscyamine 6 β -hydroxylase, a 2-oxoglutarate-dependent dioxygenase, from cultured roots of *Hyoscyamusniger*. J Biol Chem 266:9460–9464

Michael AJ, Furze JM, Rhodes MJC, Burtin D (1996) Molecular cloning and functional identification of a plant ornithine decarboxylase cDNA. Biochem J 314:241–248

Moore RC, Boyle SM (1990) Nucleotide sequence and analysis of the speA gene encoding biosynthetic arginine decarboxylase in *Escherichia coli*. J Bacterial 72:4631–4640

Moyano E, Fornalé S, Palazón J, Cusidó RM, Bagni N, Piñol MT (2002) Alkaloid production in *Duboisia* hybrid hairy root cultures overexpressing the pmt gene. Phytochemistry 59:697–702

Naik T, Vanitha SC, Rajvanshi PK, Chandrika M, Kamalraj S, Jayabaskaran C (2017) Novel microbial sources of tropane alkaloids: first report of production by endophytic fungi isolated from *Datura metel* L. Curr Microbiol 75:206–212

Nakajima K, Hashimoto T (1999) Two tropinone reductases that catalyze opposite stereospecific reductions in tropane alkaloid biosynthesis are localized in plant root with different cell-specific patterns. Plant Cell Physiol 40:1099–1107

Nakajima K, Hashimoto T, Yamada Y (1993) Two tropinone reductases with different stereospecificities are short-chain dehydrogenases evolved from a common ancestor. Proc Natl Acad Sci U S A 90:9591–9595

Nguyen TK, Jamali A, Lanoue A, Gontier E, Dauwe R (2015) Unravelling the architecture and dynamics of tropane alkaloid biosynthesis pathways using metabolite correlation networks. Phytochemistry 116:94–103

Ott SC, Tofern-Reblin B, Jenett-Siems K, Siems K, Müller F, Hilker M, Onegi B, Witte L, Eich E (2007) Unusual tropane alkaloid pattern in two African Convolvulaceous species. Phytochemistry and chemotaxonomy of the Convolvulaceae, part 20. Z Naturforsch 62:285–288

Ott SC, Jenett-Siems K, Siems K, Müller F, Hilker M, Eich E (2013) Ipvelutine, 7β-Acetoxy-2α-(tigloyloxy)tropane, an unusual tropane alkaloid from *Ipomoea velutina* R. Br. (Convolvulaceae). Sci Pharm 81:543–548

Palazón J, Moyano E, Cusidó RM, Bonfill M, Oksman-Caldentey KM, Piñol MT (2003) Alkaloid production in *Duboisia* hybrid hairy roots and plants overexpressing the *h6h* gene. Plant Sci 165:1289–1295

Paschalidis K, Tsaniklidis G, Wang BQ, Delis C, Trantas E, Loulakakis K, Makky M, Sarris PF, Ververidis F, Liu JH (2019) The interplay among polyamines and nitrogen in plant stress responses. Plan Theory 8:315

Paul A, Jha A, Bhardwaj S, Singh S, Shankar R, Kumar S (2014) RNA-seq-mediated transcriptome analysis of actively growing and winter dormant shoots identifies non-deciduous habit of evergreen tree tea during winters. Sci Rep 4:5932

Pearn J, Thearle J (1982) The history of hyoscine. Hist Sci Med 17:257–261

Poupko JM, Baskin SI, Moore E (2007) The pharmacological properties of anisodamine. J Appl Toxicol 27:116–121

Qiang W, Xia K, Zhang Q, Zeng J, Huang Y, Yang C, Chen M, Liu X, Lan X, Liao Z (2016) Functional characterisation of a tropine-forming reductase gene from *Brugmansia arborea*, a woody plant species producing tropane alkaloids. Phytochemistry 127:12–22

Qiu F, Yang C, Yuan L, Xiang D, Lan X, Chen M, Liao Z (2018) A phenylpyruvic acid reductase is required for biosynthesis of tropane alkaloids. Org Lett 20:7807–7810

Qiu F, Zeng J, Wang J, Huang JP, Zhou W, Yang C, Lan X, Chen M, Huang SX, Kai G, Liao Z (2020) Functional genomics analysis reveals two novel genes required for littorine biosynthesis. New Phytol 225:1906–1914

Richter U, Rothe G, Fabian AK, Rahfeld B, Dräger B (2005) Overexpression of tropinone reductases alters alkaloid composition in *Atropa belladonna* root cultures. J Exp Bot 56:645–652

Robins RJ, Parr AJ, Walton NJ (1991) Studies on the biosynthesis of tropane alkaloids in *Datura stramonium* L. transformed root cultures: 2. On the relative contributions of L-arginine and L-ornithine to the formation of the tropane ring. Planta 183:196–201

Rothe G, Hachiya A, Yamada Y, Hashimoto T, Dräger B (2003) Alkaloids in plants and root cultures of *Atropa belladonna* overexpressing putrescine N-methyltransferase. J Exp Bot 54:2065–2070

Sato F, Hashimoto T, Hachiya A, Tamura K, Choi KB, Morishige T, Fujimoto H, Yamada Y (2001) Metabolic engineering of plant alkaloid biosynthesis. Proc Natl Acad Sci U S A 98:367–372

Schimming T, Jenett-Siems K, Mann P, Tofern-Reblin B, Milson J, Johnson RW, Deroin T, Austin DF, Eich E (2005) Calystegines as chemotaxonomic markers in the Convolvulaceae. Phytochemistry 66:469–480

Schmidt GW, Jirschitzka J, Porta T, Reichelt M, Luck K, Torre JC, Dolke F, Varesio E, Hopfgartner G, Gershenzon J, D'Auria JC (2015) The last step in cocaine biosynthesis is catalyzed by a BAHD acyltransferase. Plant Physiol 167:89–101

Scholl Y, Höke D, Dräger B (2001) Calystegines in *Calystegia sepium* derive from the tropane alkaloid pathway. Phytochemistry 58:883–889

Senthil-Kumar M, Mysore KS (2011) New dimensions for VIGS in plant functional genomics. Trends Plant Sci 16:656–665

Shonle I, Bergelson J (2000) Evolutionary ecology of the tropane alkaloids of *Datura stramonium* L. (Solanaceae). Evolution 54:778–788

Suzuki K, Yamada Y, Hashimoto T (1999a) Expression of *Atropa belladonna* putrescine N-methyltransferase gene in root pericycle. Plant Cell Physiol 40:289–297

Suzuki K, Yun D-J, Chen X-Y, Yamada Y, Hashimoto T (1999b) An *Atropa belladonna* hyoscyamine 6β-hydroxylase gene is differentially expressed in the root pericycle and anthers. Plant Mol Biol 40:141–152

Torre JC, Schmidt GW, Paetz C, Reichelt M, Schneider B, Gershenzon J, D'Auria JC (2013) The biosynthesis of hydroxycinnamoyl quinate esters and their role in the storage of cocaine in *Erythroxylum coca*. Phytochemistry 91:177–186

Ullrich SF, Hagels H, Kayser O (2017) Scopolamine: a journey from the field to clinics. Phytochem Rev 16:333–353

Wu N, Jian D, Xiang M, Chen M, Lan X, Liao Z, Liu X (2019) Biochemical characterization reveals the functional divergence of two tropinone reductases from *Przewalskia tangutica*. Biotechnol Appl Biochem 66:597–606

Xia K, Liu X, Zhang Q, Qiang W, Guo J, Lan X, Chen M, Liao Z (2016) Promoting scopolamine biosynthesis in transgenic *Atropa belladonna* plants with pmt and h6h overexpression under field conditions. Plant Physiol Biochem 106:46–53

Zhang L, Ding R, Chai Y, Bonfill M, Moyano E, Oksman-Caldentey KM, Xu T, Pi Y, Wang Z, Zhang H, Kai G, Liao Z, Sun X, Tang K (2004) Engineering tropane biosynthetic pathway in *Hyoscyamus niger* hairy root cultures. Proc Natl Acad Sci U S A 101:6786–6791

Zhao K, Zeng J, Zhao T, Zhang H, Qiu F, Yang C, Zeng L, Liu X, Chen M, Lan X, Liao Z (2017) Enhancing tropane alkaloid production based on the functional identification of tropine-forming reductase in *Scopolia lurida*, a tibetan medicinal plant. Front Plant Sci 8:1745

Zhao T, Wang C, Bai F, Li S, Yang C, Zhang F, Bai G, Chen M, Lan X, Liao Z (2019) Metabolic characterization of *Hyoscyamus niger* ornithine decarboxylase. Front Plant Sci 10:229

Zhao T, Li S, Wang J, Zhou Q, Yang C, Bai F, Lan X, Chen M, Liao Z (2020) Engineering tropane alkaloid production based on metabolic characterization of ornithine decarboxylase in *Atropa belladonna*. ACS Synth Biol 9:437–448

Ziegler J, Facchini PJ (2008) Alkaloid biosynthesis: metabolism and trafficking. Ann Rev Plant Biol 59:735–769

Transcription Factor and MicroRNA-Mediated Manipulation of Tropane Alkaloid Biosynthesis

9

Pravin Prakash, Rakesh Srivastava, and Praveen Chandra Verma

Abstract

Transcription factors (TFs) and microRNAs (miRNAs) are considered as potent regulators of gene expression in plants. TFs and miRNAs regulate several biological processes including growth, development, response to biotic and abiotic stress conditions, as well as secondary metabolism in plants. The TF- and miRNA-mediated genetic engineering of secondary metabolic pathways is a promising approach toward the rational production of pharmaceutically and economically useful plant products. Tropane alkaloids (TAs) are commercially important secondary metabolites produced mainly by the Solanaceae and Erythroxylaceae families. Knowledge of TFs and miRNAs involved in TA biosynthesis could facilitate their utilization as potential candidates for the manipulation of the plant secondary metabolic pathways.

Keywords

Transcription factor · MicroRNA · Tropane alkaloids · Genetic manipulation · Secondary metabolism

Abbreviations

BIAs	Benzylisoquinoline alkaloids
CaMV 35S	Cauliflower mosaic virus 35S
cDNA	Complementary deoxyribonucleic acid

P. Prakash · R. Srivastava · P. C. Verma (✉)
Molecular Biology & Biotechnology Division, Council of Scientific and Industrial Research-National Botanical Research Institute (CSIR-NBRI), Lucknow, Uttar Pradesh, India
e-mail: praveencverma@nbri.res.in

V. Srivastava et al. (eds.), *Tropane Alkaloids*,
https://doi.org/10.1007/978-981-33-4535-5_9

CRISPR/Cas9	Clustered regularly interspaced short palindromic repeats/ CRISPR-associated protein 9
GC-MS	Gas chromatography-mass spectrometry
HPLC	High-performance liquid chromatography
LC-MS	Liquid chromatography-mass spectrometry
miRNA	MicroRNA
miRNA*	MicroRNA star or passenger strand
mRNA	Messenger ribonucleic acid
NCBI	National Center for Biotechnology Information
RISC	RNA-induced silencing complex
RNAi	RNA interference
sgRNA	Single-guide RNA
TALENs	Transcription activator-like effector nucleases
TAs	Tropane alkaloids
TFs	Transcription factors
TIAs	Terpenoid indole alkaloids
ZFNs	Zinc-finger nucleases

9.1 Introduction

Transcription factors (TFs) are proteins that exert their regulatory effects by interacting with the sequence-specific *cis*-elements found in the promoter regions of the genes. The DNA-binding and activation domains are the structural components of a transcription factor (Latchman 1997). TFs play a major role in the secondary metabolic pathway regulation, and several TFs have also been used to manipulate the secondary metabolic pathways in plants (Ma et al. 2009; Yang et al. 2012; Yu et al. 2012; Lu et al. 2013; Patra et al. 2013; Han et al. 2014; Tan et al. 2015; Van Moerkercke et al. 2015; Zhang et al. 2015; Shen et al. 2016; Cao et al. 2020). MicroRNAs are small, approximately 21-nucleotide (nt)-long, molecules potentially regulating the transcriptional and the post-transcriptional gene expression in plants and animals (Bartel 2004; Jones-Rhoades et al. 2006). The miRNAs generally target their complementary mRNA transcripts and regulate their expression either by cleaving or repressing them. Like TFs, several miRNAs have been shown to target the crucial genes involved in secondary metabolic pathway regulation. Tropane alkaloids (TAs) are bicyclic tropane ring system containing metabolites. These nitrogen-containing alkaloids can be categorized into three major groups: (1) cocaine, (2) calystegines, and (3) hyoscyamine and scopolamine. Cocaine is found in *Erythroxylum coca*. Calystegines are mostly found in Brassicaceae, Convolvulaceae, Erythroxylaceae, Moraceae, and Solanaceae families. Hyoscyamine and scopolamine are commonly found in the Solanaceae family of plants (Grynkiewicz and Gadzikowska 2008; Afewerki et al. 2019; Kohnen-Johannsen and Kayser 2019). Cocaine production is not legal in many countries, but still it holds the global second rank as an illicit drug in consumption (Schultze-Kraft 2016). So far, the identified calystegines have not shown to be pharmaceutically or commercially important. Hyoscyamine and scopolamine are pharmacologically active compounds mainly used as anticholinergic agents. Scopolamine is preferred over hyoscyamine as it exhibits fewer side effects (Ullrich et al. 2017). Due to their medicinal and economic importance, most TA research has been carried out to enhance the production of hyoscyamine and scopolamine. However, between these two secondary metabolites, the enhanced production of scopolamine has been largely focused by researchers due to medicinal and economic interests (Grynkiewicz and Gadzikowska 2008; Afewerki et al. 2019).

Genetic manipulation of the plant's secondary metabolic pathways requires extensive knowledge about the pathway of interest. It includes detailed information about the genes, enzymes, and regulatory factors controlling the crucial steps of the metabolic pathways. Once the important regulatory elements are identified, they can be selectively manipulated by opting for one of the several strategies available. Overexpression and downregulation/silencing of genes, TFs, and miRNAs are routinely carried out to dissect their roles in metabolic pathways. Recently, genome editing has emerged as a preferred tool for pathway manipulation. Transcription regulation is an essential step for controlling the messenger ribonucleic acid (mRNA) synthesis of a gene, which also depends on either the internal or the external environmental conditions of the cell. The production of secondary metabolites might also depend on various factors, such as the plant developmental or growth stages, signals perceived by different cell types, as well as the abiotic or the biotic stress conditions (Pavarini et al. 2012; Srivastava et al. 2014a; Pandey et al. 2016; Yang et al. 2018). These factors may directly or indirectly coordinate and control the secondary metabolite biosynthesis and accumulation through transcriptional regulation. Transcription is usually controlled at several steps by the upstream regulatory region of a gene, known as the promoter. The promoter not only initiates the transcription of a gene but also controls its expression (Singh 1998; Kiran et al. 2006; Srivastava et al. 2014b). The transcription factors interact with the *cis*-elements found in the promoters and play an important role in initiating, coordinating, and regulating the transcription process (Venters and Pugh 2009; Hahn and Young 2011; Srivastava et al. 2014b). The promoters can be classified into four major groups: constitutive (continuously expressed), inducible (expressed during particular conditions), spatiotemporal (tissue- and time-specific expression), and synthetic (artificially designed) (Srivastava et al. 2018; Pandey et al. 2019). Several studies suggest that the production of TAs could be precisely modulated by overexpressing and/or downregulating/silencing the TA biosynthesis-related potential enzyme-encoding genes under the control of a constitutive promoter (Zhang et al. 2004; Häkkinen et al. 2005; Zhao et al. 2020). Several major genes involved in the TA biosynthesis pathway have already been identified and used successfully for the TA biosynthesis pathway manipulation. In *Anisodus acutangulus*, *putrescine N-methyltransferase* (*AaPMT*) and *tropinone reductase I* (*AaTRI*) genes were co-overexpressed in the hairy root cultures. Metabolite analyses of transgenic hairy roots showed the enhanced production of four TAs, viz., anisodamine, hyoscyamine, anisodine, and scopolamine (Kai et al. 2011b). At present, the hybrid crosses between the *Duboisia myoporoides* and *Duboisia leichhardtii* plants serve as the commercial source of hyoscyamine and scopolamine. The RNA interference (RNAi)-mediated silencing of *quinolinate phosphoribosyl transferase* (*QPT*) gene led to the enhanced production of scopolamine in transgenic hairy root lines of *Duboisia leichhardtii* (Singh et al. 2018). By using the overexpression and silencing strategies, the role of *ornithine decarboxylase* (*ODC*) gene was elucidated in the TA biosynthesis in *Atropa belladonna*. The overexpression of *A. belladonna ornithine decarboxylase* (*AbODC*) led to the enhanced production of anisodamine, hyoscyamine, putrescine, N-methylputrescine, hyoscyamine, and putrescine in transgenic hairy root lines. The downregulation of *AbODC* in *A. belladonna* hairy root lines led to the reduced production of putrescine, N-methylputrescine, and other TAs (Zhao et al. 2020). The co-overexpression of genes encoding *hyoscyamine 6β-hydroxylase* (*H6H*) and *putrescine N-methyltransferase* (*PMT*) enzyme resulted in significantly higher production of scopolamine in the transgenic hairy root lines of *Hyoscyamus niger* (Zhang et al. 2004).

9.2 Transcription Factor (TF)-Mediated Regulation of Plant Secondary Metabolism: Potential in Tropane Alkaloid (TA) Biosynthesis Pathway Engineering

TFs participate in various biological processes in plants including secondary metabolism. Several TF families have been reported to participate in the regulation of plant secondary metabolic pathways. These include the *APETALA2/ethylene response factor* (*AP2/ERF*), *basic helix-loop-helix* (*bHLH*), *basic leucine zipper* (*bZIP*), DNA-binding with one finger (*DOF*), *lin-11, isl-1, mec-3* (*LIM*), *myeloblastosis* (*MYB*), *NAM, ATAF, CUC* (*NAC*), *SQUAMOSA promoter-binding protein-like* (*SPL*), *WRKY*, and Cys_2/His_2-type zinc-finger protein transcription factor families (Menke et al. 1999; Kawaoka et al. 2000; Pauw et al. 2004; Gális et al. 2006; Skirycz et al. 2006; Zheng et al. 2006; Todd et al. 2010; Gou et al. 2011; Suttipanta et al. 2011; Akagi et al. 2012; Saga et al. 2012; Yu et al. 2012). Additionally, the roles of TFs in transcriptional regulation and plant secondary metabolism have also been reviewed broadly (Yang et al. 2012; Patra et al. 2013; Zhou and Memelink 2016; Srivastava and Verma 2017). Various plant secondary metabolic pathways leading to the biosynthesis of anthocyanin, artemisinin, camalexin, flavonoid, glucosinolates, sesquiterpene, nicotine, phenylpropanoids, proanthocyanidin, terpene, and terpenoid indole alkaloids (TIAs) have been reported to be regulated by the TFs (Kawaoka et al. 2000; van der Fits and Memelink 2000; Xu et al. 2004; Gális et al. 2006; Skirycz et al. 2007; Gigolashvili et al. 2008; Todd et al. 2010; Gou et al. 2011; Suttipanta et al. 2011; Zhang et al. 2011; Saga et al. 2012; Verdier et al. 2012; Yu et al. 2012). Anthocyanin biosynthesis is reported to be regulated by the TFs belonging to the MYB, bHLH, and SPL families. Among the MYB TF family members, the MYB75/production of anthocyanin pigment1 (PAP1), MYB113, and MYB114 regulates anthocyanin biosynthesis in *Arabidopsis thaliana* (Teng et al. 2005; Gonzalez et al. 2008). *LhMYB6* and *LhMYB12* in Asiatic hybrid lily (Yamagishi et al. 2010), ruby in *Citrus sinensis* (Butelli et al. 2012), and *MdMYB1* and *MdMYB10* in *Malus domestica* (Takos et al. 2006; Espley et al. 2007) are the other MYB TF family members involved in the regulation of anthocyanin biosynthesis. The *GLABRA3* (*GL3*) and *ENHANCER OF GLABRA3* (*EGL3*) regulating the anthocyanin biosynthesis in *A. thaliana* (Feyissa et al. 2009) belong to the bHLH family of TFs. The SPL family TF, SPL9, regulates the anthocyanin biosynthesis in *A. thaliana* (Gou et al. 2011). Artemisinin biosynthesis is regulated by the AP2/ERF family TFs, *AaERF1*, and *AaERF2* in *Artemisia annua* (Yu et al. 2012). Camalexin biosynthesis is regulated by the WRKY and NAC family TFs, *WRKY33* (Zheng et al. 2006), and *ANAC042* (Saga et al. 2012), in *A. thaliana*. The DOF TF family member, *AtDOF4;2*, regulates phenylpropanoid biosynthesis in *A. thaliana* (Skirycz et al. 2007). Glucosinolate biosynthesis is known to be regulated by the MYB and DOF family TFs. Among the MYB TFs, HAG1/MYB28, HAG2/MYB76, and HAG3/MYB29 (Gigolashvili et al. 2008), ATR1 (Celenza et al. 2005), and HIG1/MYB51 (Gigolashvili et al. 2007) have been reported to regulate the glucosinolate biosynthesis in *A. thaliana*. The DOF family TF, *AtDof1.1* (*OBP2*), regulates the glucosinolate biosynthesis in *Arabidopsis* (Skirycz et al. 2006). Sesquiterpene biosynthesis has been reported to be regulated by a WRKY family TF, GaWRKY1, in cotton (Xu et al. 2004). In *Nicotiana benthamiana*, the bHLH family TFs, *NbbHLH1* and *NbbHLH2*, are reported to regulate the nicotine biosynthesis (Todd et al. 2010). In *N. tabacum*, a MYB family TF, NtMYBJS1, regulates the phenylpropanoid biosynthesis (Gális et al. 2006). Proanthocyanidin biosynthesis has been shown to be regulated by the MYB family TFs. The R2R3 MYB protein-encoding gene *TRANSPARENT TESTA2* (*TT2*) in *A. thaliana* (Nesi et al. 2001), and MtPAR, in *Medicago truncatula* (Verdier et al. 2012), regulates proanthocyanidin biosynthesis. The bZIP TF family member, DkbZIP5, regulates proanthocyanidin biosynthesis in *Diospyros kaki* (Akagi et al. 2012). The bHLH TF family member, MYC2, regulates sesquiterpene biosynthesis in *A. thaliana* (Hong et al. 2012). Terpenoid indole alkaloid (TIA) biosynthesis is regulated by the *Catharanthus roseus* AP2/ERF family TFs, ORCA2 (Menke et al. 1999) and ORCA3 (van der Fits and Memelink 2000), *C. roseus* bHLH TF CrMYC2 (Zhang et al. 2011), *C. roseus* WRKY TF family member CrWRKY1 (Suttipanta et al. 2011), and *C. roseus* zinc-finger proteins ZCT1, ZCT2, and ZCT3 (Pauw et al. 2004).

Although such studies are limited in TA-harboring plants, however, the exploration of *Anisodus acutangulus* transcriptome has indicated toward the contribution of TFs in plant secondary metabolism (Cui et al. 2015). *A. acutangulus* is a perennial plant that belongs to the Solanaceae family. The plant contains a rich amount of TAs including anisodine, hyoscyamine, and scopolamine, and it is used as a common herbal medicine in the Yunnan province of China (Wu et al. 1962; Zeng 1962; Kai et al. 2007, 2009a, b, 2011a, b, 2012). Several TF families have been identified by transcriptome sequencing and analysis in *A. acutangulus*. Transcriptome analysis in *A. acutangulus* has also revealed several potential candidate genes that belong to TF families involved in plant secondary metabolism (Cui et al. 2015). A total of 2172 TF-encoding unigenes were reported in *A. acutangulus*. Out of these, a total of

588 unigenes were found to belong to WRKY, MYB, AP2/ERF, MYC, JAZ, bHLH, and NAC TF families. Among the secondary metabolism-related TF family unigenes, MYB TF family was represented by 218 unigenes, followed by bHLH, AP2/ERF, WRKY, NAC, MYC, and JAZ TF families which were represented by 131, 120, 85, 14, 11, and 9 unigenes, respectively (Cui et al. 2015).

9.3 MicroRNAs (miRNAs) as Regulators of Plant Secondary Metabolism: Their Potential in Modulation of the Tropane Alkaloids (TA) Biosynthesis Pathway

The microRNAs (miRNAs) have recently emerged as potential candidates for plant metabolic engineering (Bulgakov and Avramenko 2015; Zhang and Wang 2015; Sabzehzari and Naghavi 2019). Several miRNAs have been identified from secondary metabolite-rich plants wherein they have been found to target the transcripts of secondary metabolic pathway genes. The prediction of key metabolic pathway genes as miRNA target indicates the possible role of these small RNAs in plant secondary metabolism regulation. There are several studies that report the miRNAs targeting the secondary metabolic pathway genes. The miRNAs are known to target secondary metabolic biosynthesis pathway genes and could be involved in the regulation of the biosynthesis of the metabolite of interest. The alkaloids, terpenoids, flavonoids, and other secondary metabolic pathway genes have been predicted to be targeted by miRNAs (Table 9.1). The miRNAs targeting various secondary metabolic pathway genes include the *Papaver somniferum* (pso)-miR13, pso-miR408, and pso-miR2161 targeting the benzylisoquinoline alkaloid (BIA) biosynthesis pathway genes in *P. somniferum* (Boke et al. 2015); *Nicotiana tabacum* (nta)-miRX17, nta-miRX19, nta-miRX20, and nta-miRX27 targeting the nicotine biosynthesis genes in *N. tabacum* (Li et al. 2015); miR156 targeting the anthocyanin biosynthesis genes in *A. thaliana* (Gou et al. 2011); miR5090 targeting the glucosinolate biosynthesis genes in *A. thaliana* (He et al. 2014); miR156 targeting the sesquiterpenoid and terpenoid biosynthesis genes in *A. thaliana* and *Pogostemon cablin* (Yu et al. 2015); *Rauvolfia serpentina* (rse)-miR396b targeting the flavonol glycoside and secologanin biosynthesis; rse-miR828a targeting the anthocyanin biosynthesis pathway genes in *R. serpentina* (Prakash et al. 2016); miR414 targeting the sesquiterpenoid and triterpenoid biosynthesis pathway genes; miR156 and miR5021 targeting the terpenoid backbone-biosynthesis genes in *Mentha* species (Singh et al. 2016a); miR-4995 targeting the picroside biosynthesis pathway genes in *Picrorhiza kurroa* (Vashisht et al. 2015); miR1134, miR5021, and miR6435 targeting the terpenoid biosynthesis genes in *Xanthium strumarium* (Fan et al. 2015); miR5072 targeting the tanshinone biosynthesis pathway genes in *Salvia miltiorrhiza* (Xu et al. 2014); *Diospyros kaki* (dka)-miR395p-3p and dka-miR858b targeting the proanthocyanidin biosynthesis pathway genes in *D. kaki* (Luo et al. 2015); *Podophyllum hexandrum* (phe)-miR172i and phe-miR1438 targeting the phenylpropanoid biosynthesis; phe-miR829.1 and phe-miR1873 targeting the flavonoid biosynthesis; phe-miR5532 targeting the isoflavonoid biosynthesis;

Table 9.1 List of miRNAs targeting the genes of various secondary metabolite biosynthesis pathways

S. no.	miRNAs	Secondary metabolic pathway	Plant species	References
1.	pso-miR13, pso-miR408, and pso-miR2161	Benzylisoquinoline alkaloid (BIA) biosynthesis	*Papaver somniferum*	Boke et al. (2015)
2.	nta-miRX17, nta-miRX19, nta-miRX20, and nta-miRX27	Nicotine biosynthesis	*Nicotiana tabacum*	Li et al. (2015)
3.	miR156	Anthocyanin biosynthesis	*Arabidopsis thaliana*	Gou et al. (2011)
4.	miR5090	Glucosinolate biosynthesis	*Arabidopsis thaliana*	He et al. (2014)
5.	miR156	Sesquiterpenoid and terpenoid	*Arabidopsis thaliana* and *Pogostemon cablin*	Yu et al. (2015)
6.	rse-miR396b	Flavonol glycoside and secologanin biosynthesis	*Rauvolfia serpentina*	Prakash et al. (2016)
7.	rse-miR828a	Anthocyanin biosynthesis	*Rauvolfia serpentina*	Prakash et al. (2016)
8.	miR414	Sesquiterpenoid and triterpenoid biosynthesis	*Mentha* species	Singh et al. (2016a)
9.	miR156 and miR5021	Terpenoid backbone biosynthesis	*Mentha* species	Singh et al. (2016a)
10.	miR-4995	Picroside biosynthesis	*Picrorhiza kurroa*	Vashisht et al. (2015)
11.	miR1134, miR5021, and miR6435	Terpenoid biosynthesis	*Xanthium strumarium*	Fan et al. (2015)
12.	miR5072	Tanshinone biosynthesis	*Salvia miltiorrhiza*	Xu et al. (2014)
13.	dka-miR395p-3p and dka-miR858b	Proanthocyanidin biosynthesis	*Diospyros kaki*	Luo et al. (2015)
14.	phe-miR172i, phe-miR1438	Phenylpropanoid biosynthesis	*Podophyllum hexandrum*	Biswas et al. (2016)
15.	phe-miR829.1, phe-miR1873	Flavonoid biosynthesis	*Podophyllum hexandrum*	Biswas et al. (2016)
16.	phe-miR5532	Isoflavonoid biosynthesis	*Podophyllum hexandrum*	Biswas et al. (2016)

(continued)

Table 9.1 (continued)

S. no.	miRNAs	Secondary metabolic pathway	Plant species	References
17.	phe-miR5021, phe-miR5538	Terpenoid backbone biosynthesis	*Podophyllum hexandrum*	Biswas et al. (2016)
18.	miR396b, miR397a, miR419, miR3711, miR5021, and miR5260	Lignin biosynthesis	*Pyrus bretschneideri*	Wu et al. (2014)
19.	miR854 and miR5021	Terpenoid backbone biosynthesis	*Zingiber officinale*	Singh et al. (2016b)
20.	miR5015b	Phenylpropanoid biosynthesis	*Zingiber officinale*	Singh et al. (2016b)

phe-miR5021 and phe-miR5538 targeting the terpenoid backbone-biosynthesis pathway genes in *P. hexandrum* (Biswas et al. 2016); miR396b, miR397a, miR419, miR3711, miR5021, and miR5260 targeting the lignin biosynthesis pathway genes in *Pyrus bretschneideri* (Wu et al. 2014); miR854 and miR5021 targeting the terpenoid backbone-biosynthesis genes; and miR5015b targeting the phenylpropanoid biosynthesis pathway gene in *Zingiber officinale* (Singh et al. 2016b).

Among the TA-producing plants, one miRNA has been identified which is involved in the cross *trans*-species and *trans*-kingdom regulation. *Atropa belladonna* (aba)-miRNA-9497 targets the zinc-finger transcription factor ZNF-691. This miRNA shared homology with *Homo sapiens* miRNA-378. The aba-miRNA-9497 has been shown to target the zinc-finger transcription factor ZNF-691 in the central nervous system (CNS) and modulated the gene expression (Avsar et al. 2020). The identification and characterization of aba-miRNA-9497 indicates toward the enormous possibilities that miRNAs identified from the TA-producing plants, besides involved in cross-kingdom regulation, might also target the genes involved in secondary metabolic pathways and regulate the biosynthesis of secondary metabolites.

9.4 Approaches for Transcription Factor (TF)- and MicroRNA (miRNA)-Mediated Pathway Modulation

9.4.1 Overexpression and Downregulation Approaches

The secondary metabolic pathways could be manipulated by TF and/or miRNA overexpression, downregulation/silencing, or editing at the genome level. The TFs generally act by binding to the *cis*-elements present in the promoter region of their target genes. This protein-DNA interaction determines the positive and/or negative regulation of their target genes. The role of a TF can be determined by modulating its expression or editing its sequence at the genomic level. This can be achieved by overexpressing, downregulating/silencing, or CRISPR-/Cas9-mediated genome editing of the TF of interest. A flow diagram depicting the use of TFs and miRNAs for TA biosynthesis manipulation is shown in Fig. 9.1. Several studies suggest that the production of the TAs could be regulated and modulated by constitutively expressing the gene of interest under the control of a suitable promoter (Zhang et al. 2004; Häkkinen et al. 2005; Zhao et al. 2020). Although the genetic modulation using the constitutive promoter has been successfully used to enhance TA production, there are also limitations with the approach, for example, the viral CaMV 35S promoter might cause the transgene-induced gene silencing (Vaucheret et al. 1998), and the high production of a product probably disturbs the plant or hairy root line growth or viability. To overcome these problems, the inducible, spatiotemporal, or artificial promoters can be used to enhance TA production. The miRNAs regulate the biological processes either by inhibiting or cleaving their target transcripts. Usually, a contrasting expression pattern is observed for the miRNAs and their target transcripts. The high expression of a particular miRNA at a given time point in a particular tissue generally leads to the prominent inhibition and thus downregulation of its target gene transcript. Similarly, the less or no expression of a miRNA in a particular tissue leads to the high expression of its target gene transcript. The role of miRNA can be checked by its *in planta* overexpression or silencing. Recently, the plant miRNAs have also been edited by using the CRISPR-/Cas9-based approach (Zhou et al. 2017). The miRNAs of interest can be overexpressed or silenced by using the suitable plant binary vectors designed for this purpose. Generally, a plant optimized miRNA precursor backbone is required to overexpress the miRNA of interest. Besides, the plant native miRNA precursor can also be cloned under the control of a CaMV 35S promoter for constitutive expression. The CaMV 35S could also be replaced by a tissue-specific promoter for the expression of miRNA in a particular tissue. For inhibiting the activity of a particular miRNA, the oligonucleotides exhibiting perfect sequence complementarity to the guide miRNA or miRNA seed region can be designed and used. These anti-miR or antagomiR oligos can be cloned in a suitable binary vector for the plant transformation.

9.4.2 Genome Editing

The CRISPR-/Cas9-mediated genome editing technique is another approach that could be utilized to target the TF or miRNA of interest for the manipulation of the plant secondary metabolic pathways. The design of an efficient single-guide RNA (sgRNA) is a prerequisite for the successful editing of the target gene. Genome editing is a highly efficient technique, which has already been applied successfully in plants. Besides plants, the use of genome editing techniques in hairy root cultures has already been demonstrated for pathway manipulation and enhanced production of secondary metabolites (Zhou et al. 2018; Zhang et al. 2020).

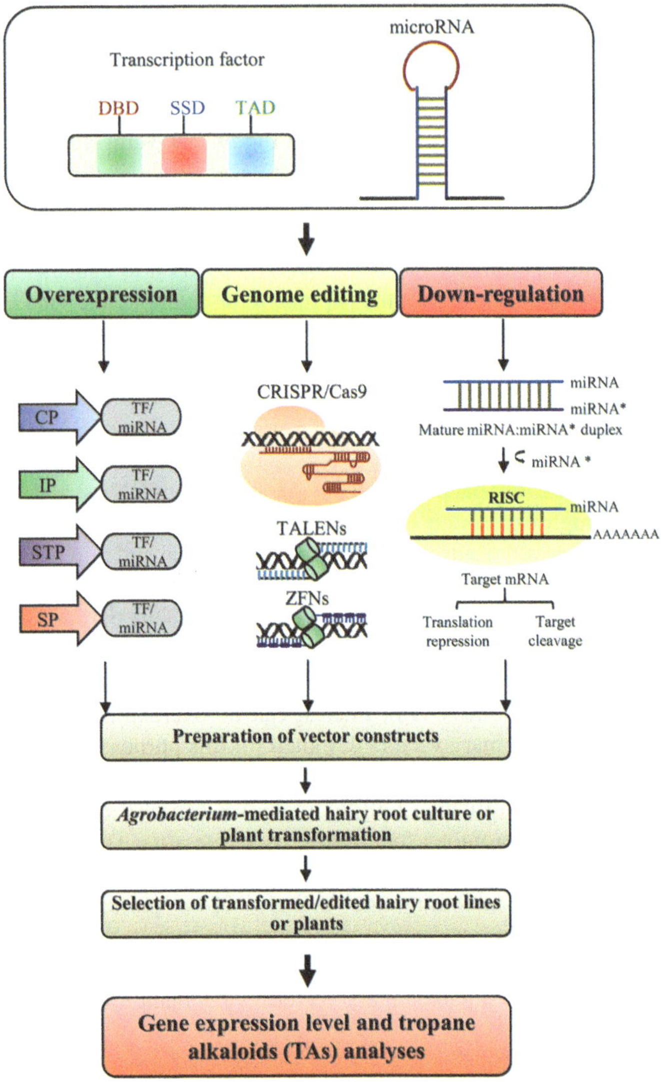

Fig. 9.1 Approaches for the transcription factor (TF)- and microRNA (miRNA)-mediated manipulation of the tropane alkaloid (TA) biosynthesis process. *TAD* trans-activating domain, *SSD* signal-sensing domain, *DBD* DNA-binding domain, *E* exon, *I* intron, *CP* constitutive promoter, *IP* inducible promoter, *STP* spatiotemporal promoter, *SP* synthetic promoter, *TF* transcription factor, *miRNA* microRNA; miRNA*, miRNA star or passenger strand, *RISC* RNA-induced silencing complex, *CRISPR/Cas9* clustered regularly interspaced short palindromic repeats/CRISPR-associated protein 9, *TALENs*, transcription activator-like effector nucleases, *ZFNs* zinc-finger nucleases

9.5 Analysis of Gene Expression Levels and Metabolite Content in Transformants

The effect of the overexpression or silencing of a TF or miRNA can be checked by measuring the expression levels of the particular TF or miRNA and its target gene. Generally, the RNA is isolated from the transgenic hairy root lines or transformed plant tissues and reverse transcribed. The complementary DNA (cDNA) is then used as template to perform the reverse transcription polymerase chain reaction (RT-PCR) or real-time PCR to quantify the gene expression levels in transformants in comparison with the controls. The indel mutations (insertion or deletions) in CRISPR-/Cas9-edited plants or hairy root lines can be reliably identified by DNA sequencing. Once the silencing/overexpression or the CRISPR-/Cas9-mediated editing of a particular TF or miRNA is confirmed, the next step is to check the effect of genetic modulation of the concerned TF or miRNA on the plant metabolites. The metabolite concentrations in the wild type and transformants could be analyzed by using the analytical techniques such as high-performance liquid chromatography (HPLC), gas chromatography-mass spectrometry (GC-MS), and liquid chromatography-mass spectrometry (LC-MS).

9.6 Conclusion

Genetic manipulation of secondary metabolic pathways has been successfully exploited for the enhanced production of commercially and pharmaceutically important metabolites from plants. TFs and miRNAs are potential gene regulators that could be utilized for the TA biosynthesis pathway manipulation. Identification of TFs and miRNAs involved in TA biosynthesis is a prerequisite for their use in pathway modulation. The in-house generated or NCBI-available genomic and transcriptomic resources of TA-producing plants could be mined for the identification of TFs and miRNAs involved in TA biosynthesis. Once the potential candidates are identified, they can be utilized for TA biosynthetic pathway manipulation. The overexpression or silencing of the TFs or miRNAs of interests has been shown to be an effective approach to decipher their roles in plant secondary metabolism. CRISPR-/Cas9-mediated editing of the TFs and/or miRNAs of interests is also gaining popularity among plant researchers. All the strategies have been proven to be effective, and the selection of any one of them by the researchers depends upon the specific strategic and experimental requirements of the concerned pathway under manipulation. TFs and miRNAs thus represent themselves as potent candidates for secondary metabolic pathway manipulation and the enhanced production of commercially and pharmaceutically important TAs in plants.

CSIR-NBRI allotted manuscript number: "**CSIR-NBRI_MS/2020/06/29**".

References

Afewerki S, Wang JX, Liao WW, Córdova A (2019) The chemical synthesis and applications of tropane alkaloids. In: The alkaloids: chemistry and biology. Academic Press, Amsterdam, pp 151–233

Akagi T, Katayama-Ikegami A, Kobayashi S, Sato A, Kono A, Yonemori K (2012) Seasonal abscisic acid signal and a basic leucine zipper transcription factor, DkbZIP5, regulate proanthocyanidin biosynthesis in persimmon fruit. Plant Physiol 158(2):1089–1102

Avsar B, Zhao Y, Li W, Lukiw WJ (2020) *Atropa belladonna* expresses a microRNA (aba-miRNA-9497) highly homologous to Homo sapiens miRNA-378 (hsa-miRNA-378); both miRNAs target the 3′-untranslated region (3′-UTR) of the mRNA encoding the neurologically relevant, zinc-finger transcription factor ZNF-691. Cell Mol Neurobiol 40(1):179–188

Bartel DP (2004) MicroRNAs: genomics, biogenesis, mechanism, and function. Cell 116 (2):281–297

Biswas S, Hazra S, Chattopadhyay S (2016) Identification of conserved miRNAs and their putative target genes in *Podophyllum hexandrum* (Himalayan Mayapple). Plant Gene 6:82–89

Boke H, Ozhuner E, Turktas M, Parmaksiz I, Ozcan S, Unver T (2015) Regulation of the alkaloid biosynthesis by miRNA in opium poppy. Plant Biotechnol J 13(3):409–420

Bulgakov VP, Avramenko TV (2015) New opportunities for the regulation of secondary metabolism in plants: focus on microRNAs. Biotechnol Lett 37(9):1719–1727

Butelli E, Licciardello C, Zhang Y, Liu J, Mackay S, Bailey P, Reforgiato-Recupero G, Martin C (2012) Retrotransposons control fruit-specific, cold-dependent accumulation of anthocyanins in blood oranges. Plant Cell 24(3):1242–1255

Cao Y, Li K, Li Y, Zhao X, Wang L (2020) MYB transcription factors as regulators of secondary metabolism in plants. Biology (Basel) 9(3):61

Celenza JL, Quiel JA, Smolen GA, Merrikh H, Silvestro AR, Normanly J, Bender J (2005) The Arabidopsis ATR1 Myb transcription factor controls indolic glucosinolate homeostasis. Plant Physiol 137(1):253–262

Cui L, Huang F, Zhang D, Lin Y, Liao P, Zong J, Kai G (2015) Transcriptome exploration for further understanding of the tropane alkaloids biosynthesis in *Anisodus acutangulus*. Mol Genet Genomics 290(4):1367–1377

Espley RV, Hellens RP, Putterill J, Stevenson DE, Kutty-Amma S, Allan AC (2007) Red colouration in apple fruit is due to the activity of the MYB transcription factor, MdMYB10. Plant J 49(3):414–427

Fan R, Li Y, Li C, Zhang Y (2015) Differential microRNA analysis of glandular trichomes and young leaves in *Xanthium strumarium* L. reveals their putative roles in regulating terpenoid biosynthesis. PLoS One 10(9):e0139002

Feyissa DN, Løvdal T, Olsen KM, Slimestad R, Lillo C (2009) The endogenous *GL3*, but not *EGL3*, gene is necessary for anthocyanin accumulation as induced by nitrogen depletion in Arabidopsis rosette stage leaves. Planta 230(4):747–754

Gális I, Šimek P, Narisawa T, Sasaki M, Horiguchi T, Fukuda H, Matsuoka K (2006) A novel R2R3 MYB transcription factor NtMYBJS1 is a methyl jasmonate-dependent regulator of phenylpropanoid-conjugate biosynthesis in tobacco. Plant J 46(4):573–592

Gigolashvili T, Berger B, Mock HP, Müller C, Weisshaar B, Flügge UI (2007) The transcription factor HIG1/MYB51 regulates indolic glucosinolate biosynthesis in *Arabidopsis thaliana*. Plant J 50(5):886–901

Gigolashvili T, Engqvist M, Yatusevich R, Müller C, Flügge UI (2008) HAG2/MYB76 and HAG3/MYB29 exert a specific and coordinated control on the regulation of aliphatic glucosinolate biosynthesis in *Arabidopsis thaliana*. New Phytol 177(3):627–642

Gonzalez A, Zhao M, Leavitt JM, Lloyd AM (2008) Regulation of the anthocyanin biosynthetic pathway by the TTG1/bHLH/Myb transcriptional complex in Arabidopsis seedlings. Plant J 53 (5):814–827

Gou JY, Felippes FF, Liu CJ, Weigel D, Wang JW (2011) Negative regulation of anthocyanin biosynthesis in Arabidopsis by a miR156-targeted SPL transcription factor. Plant Cell 23 (4):1512–1522

Grynkiewicz G, Gadzikowska M (2008) Tropane alkaloids as medicinally useful natural products and their synthetic derivatives as new drugs. Pharmacol Rep 60(4):439

Hahn S, Young ET (2011) Transcriptional regulation in *Saccharomyces cerevisiae*: transcription factor regulation and function, mechanisms of initiation, and roles of activators and coactivators. Genetics 189(3):705–736

Häkkinen ST, Moyano E, Cusido RM, Palazon J, Pinol MT, Oksman-Caldentey KM (2005) Enhanced secretion of tropane alkaloids in *Nicotiana tabacum* hairy roots expressing heterologous *hyoscyamine-6β-hydroxylase*. J Exp Bot 56(420):2611–2618

Han J, Wang H, Lundgren A, Brodelius PE (2014) Effects of overexpression of *AaWRKY1* on artemisinin biosynthesis in transgenic *Artemisia annua* plants. Phytochemistry 102:89–96

He H, Liang G, Li Y, Wang F, Yu D (2014) Two young MicroRNAs originating from target duplication mediate nitrogen starvation adaptation *via* regulation of glucosinolate synthesis in *Arabidopsis thaliana*. Plant Physiol 164(2):853–865

Hong GJ, Xue XY, Mao YB, Wang LJ, Chen XY (2012) Arabidopsis MYC2 interacts with DELLA proteins in regulating *sesquiterpene synthase* gene expression. Plant Cell 24(6):2635–2648

Jones-Rhoades MW, Bartel DP, Bartel B (2006) MicroRNAs and their regulatory roles in plants. Annu Rev Plant Biol 57:19–53

Kai GY, Chen JF, Li L, Zhou GY, Zhou LM, Zhang L, Chen YH, Zhao LX (2007) Molecular cloning and characterization of a new cDNA encoding hyoscyamine 6β-hydroxylase from roots of *Anisodus acutangulus*. J Biochem Mol Biol 40(5):715–722

Kai GY, Li L, Jiang YX, Yan XM, Zhang Y, Lu XQ, Liao P, Chen JF (2009a) Molecular cloning, characterization of two tropinone reductases in *Anisodus acutangulus* and enhancement of tropane alkaloids production in *AaTRI*-transformed hairy roots. Biotechnol Appl Biochem 54 (3):177–186

Kai GY, Zhang Y, Chen JF, Li L, Yan XM, Zhang R, Liao P, Lu X, Wang W, Zhou GY (2009b) Molecular characterization and expression analysis of two distinct putrescine N-methyltransferases from roots of *Anisodus acutangulus*. Physiol Plant 135(2):121–129

Kai GY, Xu H, Zhou CC, Liao P, Xiao JB, Luo XQ, You LJ, Zhang L (2011a) Metabolic engineering tanshinone biosynthetic pathway in *Salvia miltiorrhiza* hairy root cultures. Metab Eng 13(3):319–327

Kai G, Yang S, Luo X, Zhou W, Fu X, Zhang A, Zhang Y, Xiao J (2011b) Co-expression of *AaPMT* and *AaTRI* effectively enhances the yields of tropane alkaloids in *Anisodus acutangulus* hairy roots. BMC Biotechnol 11:43

Kai GY, Zhang A, Guo YY, Li L, Cui LJ, Luo XQ, Liu C, Xiao JB (2012) Enhancing the production of tropane alkaloids in transgenic *Anisodus acutangulus* hairy root cultures by over-expressing *tropinone reductase I* and *hyoscyamine-6β-hydroxylase*. Mol Biosyst 8 (11):2883–2890

Kawaoka A, Kaothien P, Yoshida K, Endo S, Yamada K, Ebinuma H (2000) Functional analysis of tobacco LIM protein Ntlim1 involved in lignin biosynthesis. Plant J 22(4):289–301

Kiran K, Ansari SA, Srivastava R, Lodhi N, Chaturvedi CP, Sawant SV, Tuli R (2006) The TATA-box sequence in the basal promoter contributes to determining light-dependent gene expression in plants. Plant Physiol 142(1):364–376

Kohnen-Johannsen KL, Kayser O (2019) Tropane alkaloids: chemistry, pharmacology, biosynthesis and production. Molecules 24(4):796

Latchman DS (1997) Transcription factors: an overview. Int J Biochem Cell Biol 29 (12):1305–1312

Li F, Wang W, Zhao N, Xiao B, Cao P, Wu X, Ye C, Shen E, Qiu J, Zhu QH, Xie J (2015) Regulation of nicotine biosynthesis by an endogenous target mimicry of microRNA in tobacco. Plant Physiol 169(2):1062–1071

Lu X, Zhang L, Zhang F, Jiang W, Shen Q, Zhang L, Lv Z, Wang G, Tang K (2013) AaORA, a trichome-specific AP2/ERF transcription factor of *Artemisia annua*, is a positive regulator in the artemisinin biosynthetic pathway and in disease resistance to Botrytis cinerea. New Phytol 198 (4):1191–1202

Luo Y, Zhang X, Luo Z, Zhang Q, Liu J (2015) Identification and characterization of microRNAs from Chinese pollination constant non-astringent persimmon using high-throughput sequencing. BMC Plant Biol 15(1):11

Ma D, Pu G, Lei C, Ma L, Wang H, Guo Y, Chen J, Du Z, Wang H, Li G, Ye H (2009) Isolation and characterization of AaWRKY1, an *Artemisia annua* transcription factor that regulates the *amorpha-4, 11-diene synthase* gene, a key gene of artemisinin biosynthesis. Plant Cell Physiol 50(12):2146–2161

Menke FL, Champion A, Kijne JW, Memelink J (1999) A novel jasmonate-and elicitor-responsive element in the periwinkle secondary metabolite biosynthetic gene *Str* interacts with a jasmonate- and elicitor-inducible AP2-domain transcription factor, *ORCA2*. EMBO J 18(16):4455–4463

Nesi N, Jond C, Debeaujon I, Caboche M, Lepiniec L (2001) The Arabidopsis *TT2* gene encodes an R2R3 MYB domain protein that acts as a key determinant for proanthocyanidin accumulation in developing seed. Plant Cell 13(9):2099–2114

Pandey V, Srivastava R, Akhtar N, Mishra J, Mishra P, Verma PC (2016) Expression of *Withania somnifera* steroidal glucosyltransferase gene enhances withanolide content in hairy roots. Plant Mol Biol Rep 34(3):681–689

Pandey B, Prakash P, Verma PC, Srivastava R (2019) Regulated gene expression by synthetic modulation of the promoter architecture in plants. In: Current developments in biotechnology and bioengineering. Elsevier, Amsterdam, pp 235–255

Patra B, Schluttenhofer C, Wu Y, Pattanaik S, Yuan L (2013) Transcriptional regulation of secondary metabolite biosynthesis in plants. Biochim Biophys Acta Gene Regul Mech 1829 (11):1236–1247

Pauw B, Hilliou FA, Martin VS, Chatel G, de Wolf CJ, Champion A, Pré M, van Duijn B, Kijne JW, van der Fits L, Memelink J (2004) Zinc finger proteins act as transcriptional repressors of alkaloid biosynthesis genes in *Catharanthus roseus*. J Biol Chem 279(51):52940–52948

Pavarini DP, Pavarini SP, Niehues M (2012) Lopes NP (2012) exogenous influences on plant secondary metabolite levels. Anim Feed Sci Technol 176(1–4):5–16

Prakash P, Rajakani R, Gupta V (2016) Transcriptome-wide identification of *Rauvolfia serpentina* microRNAs and prediction of their potential targets. Comput Biol Chem 61:62–74

Sabzehzari M, Naghavi MR (2019) Phyto-miRNAs-based regulation of metabolites biosynthesis in medicinal plants. Gene 682:13–24

Saga H, Ogawa T, Kai K, Suzuki H, Ogata Y, Sakurai N, Shibata D, Ohta D (2012) Identification and characterization of *ANAC042*, a transcription factor family gene involved in the regulation of camalexin biosynthesis in Arabidopsis. Mol Plant Microbe 25(5):684–696

Schultze-Kraft M (2016) Evolution of estimated coca cultivation and cocaine production in South America (Bolivia, Colombia and Peru) and of the actors, modalities and routes of cocaine trafficking to Europe; background paper commissioned by the EMCDDA for the 2016 EU drug Market. European Monitoring Centre for Drugs and Drug Addiction, Lisbon, pp 1–15

Shen Q, Lu X, Yan T, Fu X, Lv Z, Zhang F, Pan Q, Wang G, Sun X, Tang K (2016) The jasmonate-responsive AaMYC2 transcription factor positively regulates artemisinin biosynthesis in *Artemisia annua*. New Phytol 210(4):1269–1281

Singh KB (1998) Transcriptional regulation in plants: the importance of combinatorial control. Plant Physiol 118(4):1111–1120

Singh N, Srivastava S, Shasany AK, Sharma A (2016a) Identification of miRNAs and their targets involved in the secondary metabolic pathways of *Mentha* spp. Comput Biol Chem 64:154–162

Singh N, Srivastava S, Sharma A (2016b) Identification and analysis of miRNAs and their targets in ginger using bioinformatics approach. Gene 575(2):570–576

Singh P, Prasad R, Tewari R, Jaidi M, Kumar S, Rout PK, ur Rahman L (2018) Silencing of *quinolinic acid phosphoribosyl transferase* (*QPT*) gene for enhanced production of scopolamine in hairy root culture of *Duboisia leichhardtii*. Sci Rep 8(1):1–9

Skirycz A, Reichelt M, Burow M, Birkemeyer C, Rolcik J, Kopka J, Zanor MI, Gershenzon J, Strnad M, Szopa J, Mueller-Roeber B (2006) DOF transcription factor AtDof1. 1 (OBP2) is part of a regulatory network controlling glucosinolate biosynthesis in Arabidopsis. Plant J 47 (1):10–24

Skirycz A, Jozefczuk S, Stobiecki M, Muth D, Zanor MI, Witt I, Mueller-Roeber B (2007) Transcription factor AtDOF4;2 affects phenylpropanoid metabolism in *Arabidopsis thaliana*. New Phytol 175(3):425–438

Srivastava V, Verma PK (2017) The plant LIM proteins: unlocking the hidden attractions. Planta 246(3):365–375

Srivastava R, Srivastava R, Singh UM (2014a) Understanding the patterns of gene expression during climate change. In: Climate change effect on crop productivity. CRC Press, Boca Raton, pp 279–328

Srivastava R, Rai KM, Srivastava M, Kumar V, Pandey B, Singh SP, Bag SK, Singh BD, Tuli R, Sawant SV (2014b) Distinct role of core promoter architecture in regulation of light-mediated responses in plant genes. Mol Plant 7(4):626–641

Srivastava R, Rai KM, Srivastava R (2018) Plant biosynthetic engineering through transcription regulation: an insight into molecular mechanisms during environmental stress. In: Biosynthetic technology and environmental challenges. Springer, Singapore, pp 51–72

Suttipanta N, Pattanaik S, Kulshrestha M, Patra B, Singh SK, Yuan L (2011) The transcription factor CrWRKY1 positively regulates the terpenoid indole alkaloid biosynthesis in *Catharanthus roseus*. Plant Physiol 157(4):2081–2093

Takos AM, Jaffé FW, Jacob SR, Bogs J, Robinson SP, Walker AR (2006) Light-induced expression of a MYB gene regulates anthocyanin biosynthesis in red apples. Plant Physiol 142 (3):1216–1232

Tan H, Xiao L, Gao S, Li Q, Chen J, Xiao Y, Ji Q, Chen R, Chen W, Zhang L (2015) TRICHOME AND ARTEMISININ REGULATOR 1 is required for trichome development and artemisinin biosynthesis in *Artemisia annua*. Mol Plant 8(9):1396–1411

Teng S, Keurentjes J, Bentsink L, Koornneef M, Smeekens S (2005) Sucrose-specific induction of anthocyanin biosynthesis in Arabidopsis requires the *MYB75/PAP1* gene. Plant Physiol 139 (4):1840–1852

Todd AT, Liu E, Polvi SL, Pammett RT, Page JE (2010) A functional genomics screen identifies diverse transcription factors that regulate alkaloid biosynthesis in *Nicotiana benthamiana*. Plant J 62(4):589–600

Ullrich SF, Hagels H, Kayser O (2017) Scopolamine: a journey from the field to clinics. Phytochem Rev 16:333–353

van der Fits L, Memelink J (2000) ORCA3, a jasmonate-responsive transcriptional regulator of plant primary and secondary metabolism. Science 289(5477):295–297

Van Moerkercke A, Steensma P, Schweizer F, Pollier J, Gariboldi I, Payne R, Bossche RV, Miettinen K, Espoz J, Purnama PC, Kellner F (2015) The bHLH transcription factor BIS1 controls the iridoid branch of the monoterpenoid indole alkaloid pathway in *Catharanthus roseus*. Proc Natl Acad Sci U S A 112(26):8130–8135

Vashisht I, Mishra P, Pal T, Chanumolu S, Singh TR, Chauhan RS (2015) Mining NGS transcriptomes for miRNAs and dissecting their role in regulating growth, development, and secondary metabolites production in different organs of a medicinal herb, *Picrorhiza kurroa*. Planta 241(5):1255–1268

Vaucheret H, Béclin C, Elmayan T, Feuerbach F, Godon C, Morel JB, Mourrain P, Palauqui JC, Vernhettes S (1998) Transgene-induced gene silencing in plants. Plant J 16(6):651–659

Venters BJ, Pugh BF (2009) How eukaryotic genes are transcribed. Crit Rev Biochem Mol Biol 44 (2–3):117–141

Verdier J, Zhao J, Torres-Jerez I, Ge S, Liu C, He X, Mysore KS, Dixon RA, Udvardi MK (2012) MtPAR MYB transcription factor acts as an on switch for proanthocyanidin biosynthesis in *Medicago truncatula*. Proc Natl Acad Sci U S A 109(5):1766–1771

Wu DK, Wang FL, Chen ZR, Yang JS, Huang QL (1962) Chemical analysis of *Anisodus acutangulus* in Yunnan and extraction of atropine sulphate. Med Pharm Yunnan 3:67–68

Wu J, Wang D, Liu Y, Wang L, Qiao X, Zhang S (2014) Identification of miRNAs involved in pear fruit development and quality. BMC Genomics 15(1):953

Xu YH, Wang JW, Wang S, Wang JY, Chen XY (2004) Characterization of GaWRKY1, a cotton transcription factor that regulates the sesquiterpene synthase gene (+)-*δ-cadinene synthase-A*. Plant Physiol 135(1):507–515

Xu X, Jiang Q, Ma X, Ying Q, Shen B, Qian Y, Song H, Wang H (2014) Deep sequencing identifies tissue-specific microRNAs and their target genes involving in the biosynthesis of tanshinones in *Salvia miltiorrhiza*. PLoS One 9(11):e111679

Yamagishi M, Shimoyamada Y, Nakatsuka T, Masuda K (2010) Two R2R3-MYB genes, homologs of petunia *AN2*, regulate anthocyanin biosyntheses in flower tepals, tepal spots and leaves of Asiatic hybrid lily. Plant Cell Physiol 51(3):463–474

Yang CQ, Fang X, Wu XM, Mao YB, Wang LJ, Chen XY (2012) Transcriptional regulation of plant secondary metabolism. J Integr Plant Biol 54(10):703–712

Yang L, Wen KS, Ruan X, Zhao YX, Wei F, Wang Q (2018) Response of plant secondary metabolites to environmental factors. Molecules 23(4):762

Yu ZX, Li JX, Yang CQ, Hu WL, Wang LJ, Chen XY (2012) The jasmonate-responsive AP2/ERF transcription factors *AaERF1* and *AaERF2* positively regulate artemisinin biosynthesis in *Artemisia annua* L. Mol Plant 5(2):353–365

Yu ZX, Wang LJ, Zhao B, Shan CM, Zhang YH, Chen DF, Chen XY (2015) Progressive regulation of sesquiterpene biosynthesis in Arabidopsis and Patchouli (*Pogostemon cablin*) by the miR156-targeted SPL transcription factors. Mol Plant 8(1):98–110

Zeng YL (1962) Pharmaceutical study and investigation of *Anisodus acutangulus* in Yunnan. Med Pharm Yunnan 3:64–66

Zhang B, Wang Q (2015) MicroRNA-based biotechnology for plant improvement. J Cell Physiol 230(1):1–5

Zhang L, Ding R, Chai Y, Bonfill M, Moyano E, Oksman-Caldentey KM, Xu T, Pi Y, Wang Z, Zhang H, Kai G (2004) Engineering tropane biosynthetic pathway in *Hyoscyamus niger* hairy root cultures. Proc Natl Acad Sci U S A 101(17):6786–6791

Zhang H, Hedhili S, Montiel G, Zhang Y, Chatel G, Pré M, Gantet P, Memelink J (2011) The basic helix-loop-helix transcription factor CrMYC2 controls the jasmonate-responsive expression of the ORCA genes that regulate alkaloid biosynthesis in *Catharanthus roseus*. Plant J 67 (1):61–71

Zhang F, Fu X, Lv Z, Lu X, Shen Q, Zhang L, Zhu M, Wang G, Sun X, Liao Z, Tang K (2015) A basic leucine zipper transcription factor, AabZIP1, connects abscisic acid signaling with artemisinin biosynthesis in *Artemisia annua*. Mol Plant 8(1):163–175

Zhang P, Du H, Wang J, Pu Y, Yang C, Yan R, Yang H, Cheng H, Yu D (2020) Multiplex CRISPR/Cas9-mediated metabolic engineering increases soya bean isoflavone content and resistance to soya bean mosaic virus. Plant Biotechnol J 18(6):1384–1395

Zhao T, Li S, Wang J, Zhou Q, Yang C, Bai F, Lan X, Chen M, Liao Z (2020) Engineering tropane alkaloid production based on metabolic characterization of ornithine decarboxylase in *Atropa belladonna*. ACS Synth Biol 9(2):437–448

Zheng Z, Qamar SA, Chen Z, Mengiste T (2006) Arabidopsis WRKY33 transcription factor is required for resistance to necrotrophic fungal pathogens. Plant J 48(4):592–605

Zhou M, Memelink J (2016) Jasmonate-responsive transcription factors regulating plant secondary metabolism. Biotechnol Adv 34(4):441–449

Zhou J, Deng K, Cheng Y, Zhong Z, Tian L, Tang X, Tang A, Zheng X, Zhang T, Qi Y, Zhang Y (2017) CRISPR-Cas9 based genome editing reveals new insights into microRNA function and regulation in rice. Front Plant Sci 8:1598

Zhou Z, Tan H, Li Q, Chen J, Gao S, Wang Y, Chen W, Zhang L (2018) CRISPR/Cas9-mediated efficient targeted mutagenesis of RAS in *Salvia miltiorrhiza*. Phytochemistry 148:63–70

The manufacturer's authorised representative in the EU is Springer Nature Customer Service Centre GmbH, Europaplatz 3, 69115 Heidelberg, Germany. If you have any concerns regarding our products, please contact ProductSafety@springernature.com

Printed and bound by CPI Group (UK) Ltd, Croydon, CR0 4YY
30/09/2026
02225306-0020